Postphenomenological Methodologies

Postphenomenology and the Philosophy of Technology

Series Editors: Robert Rosenberger, Peter-Paul Verbeek, Don Ihde

As technologies continue to advance, they correspondingly continue to make fundamental changes to our lives. Technological changes have effects on everything from our understandings of ethics, politics, and communication, to gender, science, and selfhood. Philosophical reflection on technology can help draw out and analyze the nature of these changes, and help us to understand both the broad patterns of technological effects and the concrete details. The purpose of this series is to provide a publication outlet for field of the philosophy of technology in general, and the school of thought called "postphenomenology" in particular. The field of philosophy of technology applies insights from the history of philosophy to current issues in technology, and reflects on how technological developments change our understanding of philosophical issues. Postphenomenology is the name of an emerging research perspective used by a growing international and interdisciplinary group of scholars. This perspective utilizes insights from the philosophical tradition of phenomenology to analyze human relationships with technologies, and also integrates philosophical commitments of the American pragmatist tradition of thought.

Recent Titles in This Series

Postphenomenological Methodologies: New Ways in Mediating Techno-Human Relationships, Edited by Jesper Aagaard, Jan Kyrre Berg Friis, Jessica Sorenson, Oliver Tafdrup, and Cathrine Hasse

Postphenomenological Investigations: Essays on Human-Technology Relations, Edited by Robert Rosenberger and Peter-Paul Verbeek

Animal Constructions and Technological Knowledge by Ashley Shew

Postphenomenological Methodologies

New Ways in Mediating Techno-Human Relationships

Edited by
Jesper Aagaard, Jan Kyrre Berg Friis,
Jessica Sorenson, Oliver Tafdrup, and
Cathrine Hasse

Foreword by Don Ihde

LEXINGTON BOOKS
Lanham • Boulder • New York • London

Published by Lexington Books
An imprint of The Rowman & Littlefield Publishing Group, Inc.
4501 Forbes Boulevard, Suite 200, Lanham, Maryland 20706
www.rowman.com

Unit A, Whitacre Mews, 26-34 Stannary Street, London SE11 4AB

British Library Cataloguing in Publication Information Available

Library of Congress Cataloging-in-Publication Data Available

ISBN 978-1-4985-4523-5 (cloth : alk. paper)
ISBN 978-1-4985-4524-2 (electronic)

Postphenomenological Methodologies

New Ways in Mediating Techno-Human Relationships

Edited by
Jesper Aagaard, Jan Kyrre Berg Friis,
Jessica Sorenson, Oliver Tafdrup, and
Cathrine Hasse

Foreword by Don Ihde

LEXINGTON BOOKS
Lanham • Boulder • New York • London

Published by Lexington Books
An imprint of The Rowman & Littlefield Publishing Group, Inc.
4501 Forbes Boulevard, Suite 200, Lanham, Maryland 20706
www.rowman.com

Unit A, Whitacre Mews, 26-34 Stannary Street, London SE11 4AB

British Library Cataloguing in Publication Information Available

Library of Congress Cataloging-in-Publication Data Available

ISBN 978-1-4985-4523-5 (cloth : alk. paper)
ISBN 978-1-4985-4524-2 (electronic)

Contents

Foreword

It was obvious from its beginnings that postphenomenology was deeply interdisciplinary. It is clearly the case in this anthology, in which I see a welcome expansion of the field of postphenomenology, with researchers from many fields developing postphenomenological methodologies. My interest in human-technology relations began to take serious philosophical interest from the early 1970s, with the first articles getting to print by 1974 or 1975. One assignment I called for back then was for each student to do a daily diary of which technological objects each used during a day. Diligent students sometimes came up with a hundred or more per day! This was a very mundane way of noting that technological objects play roles in every nook and cranny of our daily lives. For students then, classroom objects were usually pens, pencils, notepads—today laptops, computer tablets, or iPads, much more to the professor's irritation. However, all this could be noticed with ordinary experience, although even that could help one become aware of one's material immersion in technologies.

Of course, classical phenomenology could sharpen this awareness, but on the subsurface, more was going on. Take the simple pencil and yellow lined writing pad: the student takes up the pencil, it feels like Merleau-Ponty's blind man's cane, the graphite touches the paper and as the pencil draws its letters and the writer feels this juncture, the letters appear as dark and narrow—very unlike the French student's dip pen with flexible nib dipped in ink—the line of the pencil remains the same width, thus just as the student "uses" the pencil, the pencil is "using" the student, demanding the same width of line throughout.

This user-used relation is two-way, reciprocal, and interactive. The writing experience is relational. The sensitive classical phenomenologist will, of course, note that this relationality is what makes up intentionality, but as suggested, in a distinctly bodily, material, Merleau-Pontean way. But also, this is a materially mediated experience—the student writes, but so does the pencil, both determining what appears on the notepad. Then, too, as Heidegger noted, the material pencil "withdraws" and becomes quasi-transparent in use. Clearly, the pencil is not the object, more the means of the writing. This is clearly not a Cartesian representation or "object" experience, but a praxical in-action experience. Much of this anthology takes precisely this praxis turn, in the field of educational experience. From blackboards to screens, the classroom technologies take their places.

Back in the time of Technics and Praxis (1979), what was new was precisely this praxis turn: modeled upon intentionality, but showing the relationality of human and technology in a relational experience. But to take materiality into intentionality, this was different, too. Much later, in Husserl's Missing Technologies (2016), I showed how Husserl retained a notion of a technology-thing, missing the dynamic in-use nature of instrument use. Something, however, remained missing in his classical phenomenology. This was the first recognition that one needed a materially sensitive post-phenomenology. A second change, however, was also underway, its beginnings showing up in the 19th century, although it did not *fully appear* until it became obvious that all science was technoscience. Indeed, I now argue that science has always been technoscience—at least as far back as the Ice Age! If science—perhaps we should call Ice Age science "archeoscience"—is the tracing and inscription of patterns of regularity in nature, then archeoscience can be found from the Ice Age. Let's take astronomy as a paradigm ancient science. Human experience, perceptual observation, is evidenced from at least 25,000 BP when humans became aware of the 28+ day lunar cycle and recorded it on reindeer antlers and inscribed stones, in many places around the world. Archeoscience evidenced. But this was also technoscience, which involved "technologies" or material implementation. For example, there was an inscription technology, the reindeer antler or inscribed stone, which made the knowledge permanent. But I argue more: there needed also to be a perceptual standardizing technology. In astronomy, this could be something as simple as a *gnomen* (stick in the ground) or, much more elaborately, a set of stones set in a circle to mark various celestial movements—and these too are ancient and are widespread in the Ice/Stone age.

Of course, this archeo-technoscience was limited; limited both by the limits of human perception, and by the then simple material instrumentation. Again, I take astronomy as my example. The primary difference between ancient astronomy and what became known as early modern science was its revolutionary technological change—optics. Post-Renaissance science saw telescopes and microscopes change human macro- and micro-perception; the new worlds of celestial and microscopic phenomena could now be perceived, and because intentionality is relational, so, too, the world expands. And, this became for me a second rationale for postphenomenology.

New technical mediations made human perception open to macro and micro phenomena never before possible. But now to push this early modern radicalization farther, one can note that in spite of the revolutionary quality of new technical mediation, all astronomy remained "eyeball optical," or limited to the ordinary bodily perceivability of "white light." This remained the case until yet again there was a "second scientific revolution" which took human perception past its classical limits of white light. The 19th century laid this groundwork when it slowly discovered what it first called "rays" which exceeded those of unmediated perception. Beginning with infrared and ultraviolet, it spread to the limits of what today is known as the electromagnetic spectrum of rays or waves—from gamma (very micro-) to radio (very macro-) which humans cannot directly, but can mediatedly experience through technologies or instruments. This second revolution is discussed in my book *Acoustic Technics* (published by Lexington, 2015). This, too, is part of the rationale for postphenomenology. As it turns out, what I am calling a second scientific revolution—clearly fully technoscientific, now in a re-doubled sense since all the phenomena revealed are revealed through mediating technologies—poses even more difficult problems for classical phenomenology, since all such phenomena lie beyond ordinary human experience, and can only be mediatedly experienced.

I must say, I was pleased to see that this anthology does avoid the temptation, which arises at the juncture of the second revolution—the temptation to fall into the non-phenomenological traps of "post-" and "trans-humanism," or presumed super-human trajectories. From my perspective, hybrid or cyborg mixes are clearly possible, but these do not slide slippery-slopishly into the super-human technofantasies of actually very ancient wishes—all of which deny contingency and mortality. I prefer an existential postphenomenology.

Mentioned in this volume are blackboards, but laptops, the Internet, digital technologies are now all common. Rather, the focus of this anthology is upon practices, and many are related to intersubjective and ethnographic practices, which long have been sensitive to postphenomenology. Several of the chapters are focused upon imaging technologies (Positron Emission Tomography (PET), telescopes, fMRI, etc.)—again, a long, deep interest of postphenomenologists. Yes, it is time for multiple disciplines to adapt postphenomenological praxes, and here we see this happening. What these novel approaches add is, maybe most of all, new questions to be dealt with empirically as well as philosophically.

I do note that since the very early days of visiting scholars and participants in the now decade-old postphenomenology research panels at the major STS conferences (begun in 2007 and continuing today), there have been many anthropologists who have added the ethnographic methods of interviews and field observations to postphenomenological practices. My contribution to this expansion of methodology is to help anthropologists become sensitive to a phenomenological way of interviewing.

The contributors to this volume bring their own senses of observation and method into play, often into dialogue with already known and published postphenomenologists, enriching the already multiple-perspectivism of postphenomenology. Just as in the established practice of presenting empirical cases in postphenomenology at STS conferences, presenters from many disciplines participate. Here I want to point out that under the strong influence of the European philosopher, Peter-Paul Verbeek, ethical considerations have been raised to major concern. Yet, one worry that so many European postphenomenologists have not (as much as our American participants) taken to heart the strong appreciation of Pragmatist emphasis, particularly that which relates to ethics. From John Dewey, to the analytic neo-pragmatism of Richard Rorty, for example, is the pragmatist resistance to what Anglo-American philosophy takes as established, a fact/value difference. Pragmatism—to which I would add postphenomenology—rejects this distinction, and this makes for a very different perspective upon ethics and materiality.

Yet, it has to be healthy for philosophy as the discipline from which postphenomenology originates, to meet and mix with the plethora of other related disciplines, such as those in this volume. It is indeed both postphenomenological and pragmatic experimentality that is called for.

Don Ihde, September 2017

An Introduction to Postphenomenological Methodologies

Jesper Aagaard, Jan Kyrre Berg Friis, Jessica Sorenson, Oliver Tafdrup, and Cathrine Hasse

The purpose of this book is to explore the idea of a *postphenomenological research methodology*: What does it mean for the field of postphenomenology to become empirically based and how might researchers go out and study human-technology relations? This discussion is an important contribution to the postphenomenological literature. Previous publications, including the Lexington book series *Postphenomenology and the Philosophy of Technology* (e.g., Friis & Crease, 2015; Wellner, 2015; Ihde, 2015; Irwin, 2016) have offered sophisticated analyses of technologies from guns and cell phones to synthesizers and earbuds, but have not dealt with the issue of methodology. Rosenberger and Verbeek's (2015) field guide does argue that postphenomenologists tend to employ case studies when studying human-technology-relations, but it also establishes that no distinctly postphenomenological methodology exists. The authors further note that the distinction between "theoretical" and "empirical" contributions may be misleading, since theoretical contributions draw on concrete examples and empirical contributions offer theoretical discussions. In this anthology, we hope to expand this point by exploring *how* empirical postphenomenologists may disclose new theoretical issues, motivate further theoretical development, and perhaps even produce novel insights. When conducting such empirical studies, researchers may be tempted to consult previous publications on phenomenological research methodologies (e.g., van Manen, 1990; Giorgi, 2009; Smith, Flowers, & Larkin, 2009). While such publications offer helpful guides to conduct phenomenological inquiry, they are often based on epistemological assumptions that run counter to key postphenomenological tenets like multistability (Ihde, 2009; Aagaard, 2017; Rosenberger, 2014; Hasse, 2008).

The present anthology is the first publication to tackle the issue of researching human-technology relations from a methodological postphenomenological perspective. Methodology refers to different things in different disciplines. In a discipline like anthropology, it refers to reflections over the methods researchers use to approach an empirical field (i.e., the sites of people's everyday lives, their lifeworlds in for instance schools or villages [e.g., Hasse, 2015]). Methods can include interviews of different kinds—participant observations, surveys, auto-ethnography. Furthermore methodology can include ethical issues tied to doing research in an empirical field. These are not separated from but rather connected to theoretical approaches. In many other disciplines, methodology refers to the analytical approach taken—for example, the analytical concepts you employ to make an analysis; in postphenomenology, these might include concepts such as multistability, variation, or mediation. As a consequence of this complexity, we have produced this book with a pressing awareness of the daunting task, but also with an exhilarating feeling of entering uncharted territory. Ultimately, if this book generates more interest in postphenomenological research and perhaps even inspires others to analyze the flaws and limitations of our approaches, we'll consider our mission to be accomplished. The goal of this anthology is not to provide cookbook recipes or to offer definitive rules for conducting postphenomenological research. Rather than providing such simple *answers*, we seek to kick-start the conversation on empirical postphenomenological research by raising some important *questions*: What does an empirically based postphenomenological investigation look like? How can it be designed? What kind of data does it produce and how might that data be analyzed? Is it always qualitative? Are qualitative interviews and participant observation equally viable? And, finally, what are the social, political, and ethical impacts of such empirically driven research?

If Heidegger has taught us one thing, however, it is that prior to raising such difficult questions, it is important to situate one's inquiry in the philosophical landscape. The next part of this introduction is devoted to this endeavor. A proper introduction to phenomenology is clearly outside the scope of this introduction, and even retracing the origins of postphenomenology is an ambitious endeavor best left to other publications (Ihde, 2009). What follows in the next outlines are a few watersheds in the transition from "classical" to "post" phenomenology that specifically focuses on methodology and concepts of use to the ensuing chapters. This narrative invariably leaves out important figures, glosses over crucial nuances, and portrays the history of phenomenology in a slightly Whiggish manner. With these caveats in mind, however, let us proceed.

INTENTIONALITY AND VARIATIONS

Husserl founded phenomenology in opposition to scientific approaches that sought to explain the world in terms of brute, physicalistic interactions. Against such naturalist reductionism, Husserl, drawing inspiration from Franz Brentano, argued that consciousness has an existence that differs radically from physical causality: Consciousness is characterized by *intentionality*, which means that it is directed toward its objects, whether real or imagined. Consciousness, in other words, is always consciousness-of-something and things appear to us only as correlates of consciousness. As a result, the task of phenomenology is to describe things exactly as they are given to consciousness. This task formed the basis of the famous motto: "To the things themselves!" What is decisive in this process, Husserl insisted (1983), is the "absolutely faithful description of what is actually present in phenomenological purity and in keeping at a distance all the interpretations transcending the given" (p. 218). To achieve such phenomenological purity, the phenomenologist must first secure freedom from prejudice by bracketing her everyday consciousness or "natural attitude." Borrowing a term from the Greek Sceptics, Husserl called this maneuver the "epoché," which means "cessation" or "suspension." After performing the epoché, the phenomenologist carefully describes an exemplary singular of a phenomenon, reduces this description to its essence, and, finally, employs a *variational analysis* that alters aspects of this preliminary definition in order to see which elements withstand such variations. Using this method, the essence of a phenomenon is abstracted from its contingent, factual features. Although Ihde (2016) has critiqued, deconstructed, and reappropriated Husserl's phenomenology in numerous ways, postphenomenology adopts Husserl's idea of intentionality as a basic directedness toward the world. Postphenomenology also employs his method of variational analysis, to which it adds a crucial twist: when employing variational analysis (e.g., on ambiguous drawings like the Necker cube), what is revealed is *multistability* rather than stable essences. Empirical studies of "things in themselves" open for questions of things in themselves *as relative to* cultural and historical diversity. These new approaches add to the Husserlian classic phenomenology approach of a human (the philosopher) investigating phenomena, by acknowledging that the moment we leave our own writing desks and mingle with people in their everyday lives, the phenomena gets complicated—as shown by many contributors to this volume. Humans stand in a variety of relations with things, not least, as emphasized in Ihde's postphenomenological explorations, relations that develop culturally and historically.

TRANSPARENCY AND BREAKDOWNS

Heidegger, Husserl's rebellious apprentice, moved us one step closer to studying things. Heidegger (1985) was highly critical of Husserl's phenomenology and argued that his notion of consciousness was drawn from philosophical tradition rather than the things themselves. This makes Husserl's natural attitude highly unnatural, because it projects a theoretical position into everyday experience, which means that "the manner in which what is experienced gives itself is defined by the feature of an objectivity for a theoretical consideration of nature, and nothing else" (Heidegger, 1985, p. 117). This objectivist attitude misconstrues our everyday being-in-the-world and is in fact "natural" among scholars (including many philosophers). In *Being and Time* (2008), Heidegger famously argued that ordinary tool use does not consist in interacting with brute physical objects, but in skillfully dealing with things and putting them to use. In such skillful coping, tools and technologies are characterized by experiential *transparency* in the sense that they withdraw from conscious awareness and allow us to focus on the task. Heidegger utilizes the now-famous concept of "ready-to-hand" (Ger. *zuhanden)* to describe this specific mode of being. However, if technological artifacts are characterized by transparency and inconspicuousness, how can we go about analyzing these elusive entities? One of Heidegger's key ideas was the concept of a *breakdown.* When coping is interrupted, he argued, the equipmental aspects of the world are lit up and brought into view. Breakdowns therefore offer fruitful opportunities to consider the ordinary roles of artifacts in everyday practices. From Heidegger, postphenomenology inherits the two notions of transparency and breakdown. Such notions prompt empirical researchers to study how technologies recede and become what Ihde (2009) calls "quasi-transparent" means, as well as how they occasionally stand forward and become salient aspects of our practices. We may also study how designers actively aspire to build this frictionless "transparency effect" into technologies, often without obtaining the expected effect in practice—as also shown by contributors to this volume.

EMBODIMENT AND HABITS

Merleau-Ponty is an even closer ally when it comes to studying people's use of material technologies. Merleau-Ponty is a peculiar figure in the postphenomenological oeuvre, however, because he is often mentioned only in passing or relegated to its footnotes (for an exception, see Hoel & Carusi, 2015). Like Heidegger, Merleau-Ponty described the basic mode of everyday life as practical and prereflective, but he further fleshed out the phenomenological account of everyday coping by emphasizing *embodiment*. According to Merleau-Ponty, the lived body is not the passive receiver of sensory stimuli or the locus of mechanical reflexes. It is a living, breathing entity that pulsates with life. It is agentic, affective, motile, and sensuous. From a first-person perspective, we *are* that body, and it is through our bodies that we are part of the world. Through repetition, our bodies become so familiar with performing certain activities that this performance eventually operates below the threshold of awareness. This is what Merleau-Ponty calls a *habit*. Whenever we acquire such habits, they "unlock" new affordances for us by disposing us to perceive and act in certain ways. Merleau-Ponty (2002) gave various examples of this phenomenon, such as a woman automatically dodging doorframes when wearing a feathered hat, a blind man skillfully using his stick to navigate, and a driver effortlessly parking his car. Accordingly, habits blur the classical distinction between subject and object, between body and world: "Habit expresses our power of dilating our being-in-the-world, or changing our existence by appropriating fresh instruments" (Merleau-Ponty, 2002, p. 166). The body is not a fixed entity, but a fleshy network that incorporates and extends over material artifacts. Accepting this argument urges empirical researchers to study different kinds of embodiments, including Ihde's (2002) "body one" and "body two"—the lived and the cultural body—and exploring bio-cultural boundaries and collectivities.

MULTISTABILITY AND MEDIATIONS

Ihde developed postphenomenology in order to give credence to the material artifacts of our world. As Ihde (1998) puts it: "The postmodern hermeneutics of things must find ways to give voices to things, to let them speak for themselves" (p. 158). This idea of letting things speak for themselves, however, does not mean that postphenomenology is an "empty" theory. On the contrary, it contains several substantial commitments: in its exploration of how we handle and incorporate technologies in our everyday practices, postphenomenology adheres to bodily experience with material technologies, and is grounded in a relational ontology that eschews any transcendental claims (Ihde, 1990). It also adheres closely to the two concepts of multistability and mediation. *Multistability* signifies how even the simplest technology has no singular essence, but can be taken up for different purposes or *stabilities* in different contexts. A lighter, for instance, is usually used to light candles or cigarettes, but can also be used to open bottles. There is no "essential" use of the lighter. *Mediation*, on the other hand, designates how technologies actively shape the relation between humans and their world. As an example, Ihde (1979) describes how a dentist's use of a sickle probe amplifies features like texture and hardness, but reduces features like moisture and temperature. Technologies not only transform perceptions, however, they also invite (Verbeek, 2005) and facilitate (Rosenberger, 2014) certain actions, while inhibiting and foreclosing others. Accordingly, they do not afford action possibilities to preexisting subjects with fixed goals, but subtly guide, nudge, and steer our intentionality (Verbeek, 2005). Combined, these concepts help us avoid the pitfalls of determinism and instrumentalism. "Technology is never purely determinative," Verbeek (2005) argues, "for in principle other cultural relations with a given artifact are always possible. But neither is it purely instrumental, for when an artifact receives a particular definition within a cultural context—and thus becomes stable rather than multistable—it still contributes to shaping that context" (p. 138). While technologies are always multistable, every stability affects the human-world relation. Postphenomenology's task is to find out how.

TAKING THE EMPIRICAL TURN

Postphenomenology has an ambivalent relationship to Heidegger's philosophy of technology. On the one hand, Ihde's (1990) widely cited typology of human-technology relations explicitly builds on Heidegger's (2008) early tool-analysis. On the other hand, postphenomenology has deliberately distanced itself from Heidegger's (1977) later description of Technology (transcendentalized with a capital T) as a unified and monolithic entity that reveals every aspect of the world—including ourselves—as resources to be optimized and utilized with maximal efficiency. The problem with this approach is not that it is wrong per se—indeed, as an interpretation of a large-scale pattern, it might be alarmingly accurate—but that the things themselves cease to matter. In other words, if our analysis departs from Heidegger's discourse on the essence of Technology, we are bound to conclude that a given artifact discloses reality as a stockpile of resources. In this case, Technology may speak through the artifact, but the artifact itself remains silent and mute. Its contribution to the world vanishes. Ihde (2006) describes his aversion to this explanatory model accordingly: "As a pragmatist and a rigorous phenomenologist, I realized this meant, simply, that such an analysis was useless since it could not discriminate between the results of playing a musical instrument, also a technological mediation, and the process of genetic manipulation!" (p. 271). Unsatisfied with this state of affairs, postphenomenology inverts or at least symmetrizes Heidegger's ontological difference, with which he famously argued that Being cannot be explained through entities: what things do equally cannot be explained with recourse to the abstract being of Technology (Verbeek, 2005). Accordingly, postphenomenology breaks with Heidegger's "one-size-fits-all" account of Technology (Ihde, 2010) and replaces his high-altitude approach with close-up studies of particular technologies—in other words, it belongs to *the empirical turn* in the philosophy of technology (Achterhuis, 2001).

METHODOLOGY AND EMPIRICAL DATA

The empirical turn has long been a contentious issue. Its proponents are often criticized for neglecting the wider, ontological atmosphere in which particular technologies are situated (for a discussion of this critique, see Rosenberger, 2017). Our issue with postphenomenology, however, is not that it is too empirical, but that it is not empirical enough. Postphenomenologists often base their analyses on texts from science journals and magazines or from their own personal life stories. While such auto-ethnographical explorations may be perfectly adequate for dealing with common household items or public places, what about practices in which the researcher is a stranger (e.g., schools, hospitals, or businesses)? As it stands, researchers who want to study such "foreign" fields must find methodological guidance elsewhere. Taken to its extreme, this dooms postphenomenology to study only what it knows, which leaves it open to Mol's (2010) caricature of phenomenology as a science that "elevates a single person's self-ethnography to grandiose proportions" (p. 254). Perhaps postphenomenology's reliance on auto-ethnographical examples has shaped (and restricted) its framework? After all, postphenomenology is chiefly concerned with how singular bodies are orientated by and toward technologies. "In a postphenomenological perspective, technology is often seen from the position of the individual rather than the collective body" (Hasse, 2015, p. 281). In contrast, many of the empirical contributions in this book emphasize the social and cultural aspects of human experience. Perhaps experiences from the *empirical field* may improve the *analytical field* of postphenomenology (Hasse, 2015). In other words, using the empirical methods of social science may yield theoretical benefits.

OVERVIEW OF THE PRESENT VOLUME

Postphenomenology, which started as a branch of philosophy, is slowly converging with social sciences like sociology, psychology, and anthropology. Conversely, postphenomenology's distinctly philosophical way of thinking about and analyzing technologies is making its mark in these empirically based disciplines. It is our contention that all research conducted under the banner of postphenomenology must meet two criteria: it must be anchored in an anti-essentialist, relational ontology (the "post-" part) and it must take departure in embodied experience (the "phenomenology" part). Apart from that, the field is wide open. None of the editors of this book are postphenomenological purists, and we believe that a certain amount of cross-pollination of intellectual vocabularies is very fertile when letting a thousand methodological flowers bloom. Accordingly, no contributor had to sign an oath of allegiance to participate in this anthology and several chapters draw on external theories and traditions to enhance their analyses. Nevertheless, all chapters somehow put postphenomenological concepts like multistability and mediation to work on technologies from blackboards and skatestoppers to robots and brain scanners. In doing so, they help to expand and clarify the postphenomenological framework and outline tentative answers to the methodological questions posed above.

The volume is structured from the principle of relationships following Ihde's I-technology-world relations (1990)—but we have given technology prevalence in the ordering of contributions.

Educational Technologies

In this first section, the authors look at *educational technologies,* such as blackboards and laptops. These authors all engage in fieldwork studies and take up important postphenomenological concepts. Variation is a focus in Catherine Adams and Joni Turville's review of postphenomenological methodologies in education research. Tobias Röhl examines transparency in educational technologies, drawing on the concepts of alterity relations and mediation. And, Jesper Aagaard considers the effects digital mediating technologies have on attention and examines these relations through the conceptual framework of multistability.

In Chapter 1, Catherine Adams and Joni Turville propose a postphenomenology of practice as a way to do postphenomenological research in education. After introducing the basic tenets of postphenomenology, Adams and Turville review existing phenomenological studies of educational technologies, tracing Ihde's influence on the field to as far back as 1975. Taking departure from these investigations and providing rich empirical examples, the chapter proceeds to outline five methodological heuristics that may help researchers wishing to pursue a postphenomenological study: (1) conducting phenomenologically sensitive, auto-ethnographic explorations; (2) observing others in action with their technologies; (3) eliciting lived experience descriptions through interview; (4) employing phenomenology's epoché-reduction, a methodological complex aimed at restoring a thing to its lived, relational wholeness; and finally (5) studying breakdowns and using "broken hammers" (and hands!) to notice what is usually taken for granted.

In Chapter 2, Tobias Röhl draws on postphenomenological concepts to account for the materiality of classroom interaction. By adopting such a socio-material view on education, two types of educational technologies can be identified: transparent media like blackboards and epistemic objects like experimental set-ups. Transparent media recede into the background and thereby allow other material to become the thematic foreground, whereas experimental set-ups become the focal object of attention. Röhl then distills the methodological consequences of his research into both these kinds of media and introduces four postphenomenological methods of analysis: (1) maximally contrasting artifacts; (2) minimally contrasting artifacts; (3) contrasting their contexts of use (e.g., the blackboard as an artifact upon which to write and behind which to hide); and (4) auto-ethnographic observation, which entails "slackening ones intentional threads" and forcing a gestalt switch that moves the artifact from taken-for-granted background to thematic foreground.

In Chapter 3, Jesper Aagaard presents a postphenomenological study of digital mediation and attention. Aagaard sets out by making Gallit Wellner's *wall-window* metaphor subject for a critical discussion. The wall-window metaphor suggests that digital technologies create an imaginary wall-window between you and the person/object you engage with, dislocating your attention to a virtual space, while still allowing you to be bodily engaged in the physical space. Aagaard criticizes this "non-reductive attentional split," by emphasizing his empirical findings from a Danish business school. This leads him to propose *the portal* as a metaphor better suited for capturing how digital mediations affect attention. He suggests that the screen functions as a portal through which the user enters another realm. Based on his empirical data, he argues that digital mediation renders the user both absent and present at the same time. Thus, he argues, the user's attention is significantly affected by digital mediation. Aagaard then proceeds to discuss the methodological concepts of reflexivity and validity in relation to his own postphenomenological study.

Self-Tracking and Imaging Technologies

In this section, the focus is on *self-tracking and imaging technologies*. Here, Moa Petersén looks at biohacking and self-tracking through a fieldwork approach that makes her question assumptions of embodiment. Fernando Secomandi also looks at self-tracking technologies, using ethnographic methods, and discusses how empirical investigation sheds light on the intersubjective constitution of human-technology relations. Finally, Ciano Aydin studies brain-scanning technologies, like functional Magnetic Resonance Imaging (fMRI). Drawing upon "technical mediation theory," Aydin investigates how brain imaging technologies mediate the conception of the brain as the locus of the self.

In Chapter 4, Moa Petersén takes up three important methodological problems for postphenomenological methodology that have emerged as issues in relation to her fieldwork in a Swedish biohacking community. The first concern is how to make the transition from a larger knowledge-producing community, like a network of biohackers to individuals. The next problem begins with an acknowledgment of Don Ihde's human-technology relations. These relations do not cover what Petersén identifies as a non-relation of a "body in a biomedial collapse." The final problem concerns the validity of a study that moves from the wider networked community to the technology-infused body. The emphasis on multistability that destabilizes the apparently homogenous technological entity, poses a threat to any claims of validity of postphenomenological methodology—apart from a focus on the multistability itself. This approach does not account for the systematicity found at a network level. Petersén suggests that that these challenges can be overcome by a cyclic process with an emphasis on studies of multistabilities found at the individual level that act as a deconstructor of the study at the network level, where we find an apparent stability and systematicity. It is the strength of postphenomenology, it is argued, that allows for these movements; and this postphenomenological methodology is needed within the social and cultural sciences.

In Chapter 5, Fernando Secomandi studies self-tracking technologies through in-depth interviews, observations, and Internet studies—and demonstrates the value of empirical research methods to postphenomenology. Postphenomenology is regularly lauded by philosophers of technology for providing impetus to a recent "empirical turn" in the field. From a philosophical perspective, to be empirical means, among other things, distancing oneself from sweeping analyses at high abstraction levels, in order to concentrate on the particularities of real-world practices of technology design and use. Secomandi argues that an empirical investigation of a self-tracking technology, inspired by an ethnographic approach to the study of service design practices, results in the acknowledgment of an overlooked topic in postphenomenology: the intersubjective constitution of human-technology relations.

In Chapter 6, Ciano Aydin examines brain imaging technologies, such as Positron Emission Tomography (PET) and functional Magnetic Resonance Imaging (fMRI), which are increasingly used not only to diagnose diseases and lesions but also to correlate brain activation with psychological states and traits. Drawing upon "technical mediation theory," Aydin investigates how brain imaging technologies mediate the conception of the brain as the locus of the self, which is then granted or denied free will. This mediated notion of self and freedom has, according to him, a history that can be traced back to Descartes's adaptation of a particular technology: the camera obscura. The camera obscura has inspired Descartes's view of the autonomous self, and is sustained and transformed by brain imaging technologies that scientists use today to prove or disprove the existence of free will. An anthropology is offered that proposes a different interpretation of the images displayed by brain imaging technologies, which also has the potential to reframe "free will" debates.

Robotic Technologies

In this section, the focus is on robotic technologies. Michael Funk discusses the theoretical benefits of comparing modern social robot engineering with the praxis of paleoanthropology from the perspective of the postphenomenological concept of material hermeneutics. Lasse Blond and Kasper Schiølin introduce notions of "stable worlds" and "unstable humans" to Ihde's *I-technology-world* approach, in their discussion of technology transfer and multistability, illustrated by the journey of the robot Silbot across multiple sites (South Korea, Denmark, and Finland).

In Chapter 7, Michael Funk discusses the methodological analogies between paleoanthropology and social robot engineering. He argues that alterity relations play an important methodological part in the paleoanthropological investigation of deep human history. Through the concept of *observer´s self-alterity,* Funk argues that paleoanthropology is not just a material hermeneutic practice of studying fossils, but also a storytelling practice that is dependent upon the ability of the researcher to conduct different interpretations of human evolution and cultural development. Funk proceeds by comparing this methodological claim with robot engineering. He argues that a similar kind of observer´s self-alterity is present in technoscientific practices of engineering social robots. The construction of social robots implies an interpretation of social practices and thus an interpretation of the involved human characteristics. Funk concludes with seven, also heuristic and schematic, theses.

In Chapter 8, Lasse Blond and Kasper Schiølin address the neglected role of humans and their lifeworlds in postphenomenology's famous *I-technology-world* formula. To expand this focus, Blond and Schiølin introduce three concepts. First, *technological style* is the everyday appropriation of technologies, which depends on human characteristics like health, age, and even mood. Second, *human instability* highlights how the configuration of such characteristics is inherently multistable. Finally, *apparently stable world* refers to the cultural context in which humans are inextricably situated. This is illustrated by an empirical study of the socially assistive robot Silbot's challenging journey from South Korean schools to Finnish and Danish nursing homes and rehabilitation centers, which eventually lead to the robot being culturally shaped in order to embed it into local practice.

General Methodological Issues

In this final section, we present general methodological discussions. Robert Rosenberger explores the theoretical benefits of amalgamating postphenomenology with other STS perspectives, like actor-network theory (ANT) and social construction of technology (SCOT). His exploration of park benches in "LOVE Park" includes a cross-variational approach to studies of particular park benches, and draws in the postphenomenological concept of multistability. Michel Puech discusses the ethical strength of postphenomenological methods, making use of illustrative case study of data immersion and exploring the postphenomenological concept of mediation. Arun Tripathi, on the other hand, writes about how the multistability of technological embodiments is characterized in order to visualize multistable hermeneutic relations with the use of technologies. Finally, Cathrine Hasse, drawing on an empirical study of participant observation among physics students, discusses the relation between the postphenomenological concepts of mediation and multistability.

In Chapter 9, Robert Rosenberger utilizes the ongoing dispute between skateboarders and anti-skateboarding groups as a case, to argue for the need of combining STS approaches like ANT and SCOT with postphenomenology. The dispute is rooted in two opposing attitudes toward the use of public space. The skateboarders represent an explorative and creative use of public space, and the anti-skateboarding groups represent an attitude of regularity and order—that materialize in different anti-skateboarding devices like "skatestoppers" built into the architecture of public space. Rosenberger argues that neither postphenomenology nor other STS perspectives alone are capable of fully grasping the complexity of the skateboarding case. He presents three arguments for an amalgamation of the two schools: (1) *The argument from interface* that claims ANT can benefit from the postphenomenological insights into how technology mediates human experience; (2) *the argument from multistability* that claims that the postphenomenological conception of materiality as multistable enables an understanding of the multiple uses of technological artifacts; and (3) *the argument from inscription*, that suggests that postphenomenology can benefit from the insights offered by ANT into how technological artifacts are inscribed into social networks.

In Chapter 10, Michel Puech examines the ethical strength of postphenomenological methods and identifies three principles for a constructive technoethics. These principles help illuminate what it means to assess what Puech describes as an "ordinary technological mediation" between a human interface, a self, and a collective value system tied to what he terms the "technosphere," which has a tendency to remain hidden. The first principle is the individual pragmatism where assessing ordinary technological mediation rests on the classical epoché—a descriptive method that opens for a valuation of usually neglected mediation. The second principle is "flourishing," making use of situated descriptions of how self, institutional politics, and practical wisdom merge. The last principle is informed by Asian philosophy and involves an integration of the global that is tied to micro-phenomenology through a search for harmony and balance, in order to explore the technosphere in its ontological, existential, and ethical specificity. The argument is underpinned by the mundane descriptions of energy consumption and data immersion in ways that make it possible to examine and assess the ordinary in our mediation with these examples from the technosphere. As there is always already meaning and intentionality in our mediation with the world, there is always a potential value-attitude in all these ordinary intentionalities. Bridging the descriptive/normative gap naturally has deep motives in the ancestry of postphenomenology.

In Chapter 11, Arun Tripathi delves into specific practices in postphenomenological methodologies and how these practices and concepts are integral to analysis. Understanding human-technology relationships has become one of the most addressed activities within philosophy of technology. The main purpose of this chapter is discussing the importance of postphenomenological methodology concerning technological mediations, focusing on "cultural variability" of technological mediations. Tripathi explores the hermeneutical practices concerning human embodiments in order to understand what technologies embody; in his chapter he discusses the question of how multistability of technological embodiments are characterized in order to visualize multistable hermeneutic relations with the use of technologies.

In Chapter 12, Cathrine Hasse introduces participant observation as the methodological foundation for an empirically grounded postphenomenology. Postphenomenology has been long conceived of itself as a move from studying "Technology" to analyzing specific technologies—the so-called *empirical turn*. To this credo, Hasse argues that postphenomenologists must also begin to look at *other* people's technological mediated practices and advocates participant observation for this purpose. As an example, she describes her own field study of physics students in which students were taught to operate an obsolete telescope as a rite of passage. Hasse further describes how students would align their measurements to avoid looking stupid—an insight that gives us a glimpse of some important social (or "non-functional") dimensions of technology use, like identity formation and pride.

REFERENCES

Aagaard, J. (2017). Introducing postphenomenological research: A brief and selective sketch of phenomenological research methods. *International Journal of Qualitative Studies in Education*, *30*(6), 519–533.

Achterhuis, H. (2001). *American Philosophy of Technology: The Empirical Turn* (R. Crease, Trans.). Bloomington, IN: Indiana University Press.

Adams, C., & Thompson, T. L. (2016). *Researching a Posthuman World: Interviews with Digital Objects*. London: Palgrave Macmillan.

Forss, A. (2012). Cells and the (imaginary) patient: The multistable practitioner-technology-cell interface in the cytology laboratory. *Medicine, Health Care and Philosophy*, *15*(3), 295–308.

Friis, J. K. B., & Crease, R. (2015). *Technoscience and Postphenomenology: The Manhattan Papers*. Lanham, MD: Lexington Books.

Giorgi, A. (2009). *The Descriptive Phenomenological Method in Psychology: A Modified Husserlian Approach*. Pittsburgh, PA: Duquesne University Press.

Hasse, C. (2008). Postphenomenology: Learning cultural perception in science. *Human Studies*, *31*(1), 43.

———. (2015). *An Anthropology of Learning: On Nested Frictions in Cultural Ecologies*. Dordrecht: Springer.

Heidegger, M. (1977). *The Question Concerning Technology and other Essays*. New York, NY: Harper & Row.

———. (1985). *History of the Concept of Time: Prolegomena*. Bloomington, IN: Indiana University Press.

———. (2008). *Being and Time*. New York, NY: Harper Perennial.

Hoel, A. S., & Carusi, A. (2015). Thinking technology with Merleau-Ponty. In R. Rosenberger & P.-P. Verbeek (Eds.), *Postphenomenological Investigations: Essays on Human-Technology Relations* (pp. 73–84). Lanham, MD: Lexington Books/Rowman & Littlefield Press.

Husserl, E. (1983). *Ideas Pertaining to a Pure Phenomenology and to a Phenomenological Philosophy. First Book*. Dordrecht: Kluwer Academic Publishers.

Ihde, D. (1979). *Technics and Praxis*. Dordrecht: Reidel Publishing.

———. (1990). *Technology and the Lifeworld: From Garden to Earth*. Indianapolis, IN: Indiana University Press.

———. (1998). *Expanding Hermeneutics: Visualism in Science*. Evanston, IL: Northwestern University Press.

———. (2002). *Bodies in Technology*. Minneapolis, MN: University of Minnesota Press.

———. (2006). Forty years in the wilderness. In E. Selinger (Ed.), *Postphenomenology: A Critical Companion to Ihde* (pp. 267–290). New York, NY: SUNY Press.

———. (2009). *Postphenomenology and Technoscience: The Peking University Lectures*. Albany, NY: State University of New York Press.

———. (2010). *Heidegger's Technologies: Postphenomenological Perspectives*. New York, NY: Fordham University Press.

———. (2015). *Acoustic Technics*. Lanham, MD: Lexington Books.

———. (2016). *Husserl's Missing Technologies*. New York, NY: Fordham University Press.

Irwin, S. (2016). *Digital Media: Human-Technology Connection*. Lanham, MD: Lexington Books.

Merleau-Ponty, M. (2002). *Phenomenology of Perception*. London: Routledge Classics.

Mol, A. (2010). Actor-Network Theory: Sensitive terms and enduring tensions. *Kölner Zeitschrift für Soziologie und Sozialpsychologie*, *50*(1), 253–269.

Rosenberger, R. (2014). Multistability and the agency of mundane artifacts: From speed bumps to subway benches. *Human Studies, 37*(3), 369–392.

———. (2017). Notes on a nonfoundational phenomenology of technology. *Foundations of Science*, *22*(3), 471–494.

Rosenberger, R., & Verbeek, P.-P. (2015). A field guide to postphenomenology. In R. Rosenberger & P.-P. Verbeek (Eds.), *Postphenomenological Investigations: Essays on Human-Technology Relations* (pp. 9–41). Lanham, MD: Lexington Books/Rowman Littlefield Press.

Smith, J., Flowers, P., & Larkin, M. (2009). *Interpretative Phenomenological Analysis: Theory, Method and Research*. Los Angeles, CA: Sage Publications.

van Manen, M. (1990). *Researching Lived Experience: Human Science for an Action Sensitive Pedagogy*. Albany, NY: SUNY Press.

Verbeek, P.-P. (2005). *What Things Do: Philosophical Reflections on Technology, Agency, and Design*. University Park, PA: Pennsylvania State University Press.

Wellner, G. (2015). *A Postphenomenological Inquiry of Cell Phones: Genealogies, Meanings, and Becoming*. Lanham, MD: Lexington Books.

Part I

EDUCATIONAL TECHNOLOGIES

Chapter 1

Doing Postphenomenology in Education

Catherine Adams and Joni Turville

INTRODUCTION

Today's classrooms—online, hybrid, flipped, one-to-one laptop—represent significant shifts in how teaching and learning transpires. Yet how do we account for these reconfigurations and future transformations? For decades, educational researchers have been attempting to articulate the positive as well as negative *differences* that digital technologies may be making to teaching, to learning, and to knowledge creation. One qualitative way to approach this question concerning technology is via phenomenology:

> Although digital media are fairly recent phenomena, it could be said that the phenomenological movement, from its earliest stages on, has offered a most fruitful starting point for understanding the meaning and the impact of these media for our lives, in general, and for education, in particular. (Vlieghe, 2015, p. 1)

Indeed, over 40 years ago, Don Ihde (1975) forwarded phenomenology as a way to interrogate "the learning process vis-à-vis machines and in terms of direct connections between man-machine relations" (p. 201). In much educational technology research today, there remains an assumption that technology either enhances learning in and of itself or that it is neutral without teacher intervention; as such, little has been done to research technologies in terms of *both* their affordances and limitations in educational lifeworlds (Aagaard, 2015a; Adams, 2010; Adams & Pente, 2011). Here, phenomenology and its technology-focused offspring postphenomenology offer promising qualitative approaches for unearthing how *specific* technologies may be reforming and transforming experiences and knowledge construction in education.

Drawing on examples from our own and others' empirical studies of digital technologies across a variety of teaching and learning environments, in this chapter we focus on *doing* postphenomenological research in education. We define postphenomenology as phenomenology that attends to specific technologies and the existential and epistemological differences they may be making to the lifeworld. We begin with a brief introduction to postphenomenology, its historical roots and associated methods. We review the status of postphenomenology in key educational technology research studies to date. Finally, we offer a set of methodological heuristics that we have found helpful in conducting our own postphenomenological inquiries of digitally textured educational lifeworlds.

Our contention is that *doing* postphenomenology must maintain a significant alliance with its progenitor, phenomenology. As Ash and Simpson (2014) recently put it, "Post-phenomenology is not about abandoning the key insights of phenomenology . . . [but rather] about refiguring and expanding phenomenology's analytic and conceptual boundaries [to] an excessive world that lies outside of the human-environment correlate but which is central to shaping human capacities, relations and experiences" (p. 16). To this end, we identify several human science methods (i.e., composing anecdotes through self-observation, gathering lived experience descriptions via interview, and observation of humans and technologies in action), as well as one philosophical method (i.e., studying breakdowns as one way to perform variational analysis or phenomenology's epoché-reduction), that we believe may be fruitfully employed in conducting this kind of inquiry. We separate these methods into two categories, roughly equivalent to data collection and data analysis in qualitative research: (1) collecting or generating *prereflective* materials or anecdotes, and (2) *reflecting* phenomenologically on the gathered prereflective materials. These disciplined enactments of human science inquiry extend a particular form of qualitative research called "phenomenology of practice" (Adams & van Manen, 2008; van Manen, 2014) to the study of technologies in educational settings. Phenomenology of practice engages many of the core gestures of both phenomenology and postphenomenology, and too, aligns with Ihde's empirical turn in the investigation of human-technology-world relations. As other postphenomenologists have done, we also leverage insights and practical understandings gleaned from Latour's actor-network theory (ANT) during the reflective phases of our analyses. We tentatively call this form of educational research a *post*phenomenology of practice.

POSTPHENOMENOLOGY AND ITS METHODS

In *Husserl's Missing Technologies*, Don Ihde (2016) describes his postphenomenology as having evolved along several lines of inquiry in North

America and Europe. The main stream established its bedrock at Ihde's home institution, Stony Brook University in the 1990s, by advancing his own pragmatic phenomenology of technics (1990), and through reading seminars on science-technology studies (STS) literature including the work of Donna Haraway, Bruno Latour, and Andrew Pickering. A second tributary has flowed out of the work of Peter-Paul Verbeek—a student of Hans Achterhuis and Ihde—and his research group at University of Twente. In both instances, postphenomenology has shared significant dialogues with science and technology studies (STS) and Latour's ANT. However, it was Verbeek's *What Things Do* (2005) that most explicitly established the amalgam of postphenomenology with ANT.

Generally, speaking, postphenomenology and ANT seem to play well together as complementary methodologies in revealing a more inclusive account of technologies (Adams & Thompson, 2017; Rosenberger, 2014). Meanwhile, Ihde seldom evokes ANT in his inquiries, and Latour has been openly antagonistic toward phenomenology. Nonetheless, because of ANT's ongoing ties to postphenomenology since Verbeek (2005), ANT necessarily figures prominently in some of the educational research discussions below. In this chapter, however, our primary focus is on the methodology of postphenomenology as an offshoot of phenomenology, and less on its relationship to ANT.

Ihde (2016) admits that his phenomenological approach to studying technologies began as the classic Husserlian variety, and that his early inquiries led him unexpectedly into territory already occupied by Martin Heidegger and Maurice Merleau-Ponty. For example, in his empirical investigations, he quickly realized that Husserl's notion of intentionality needed a radical revision, a change that turned out to be more in line with Heidegger's (1962) being-in-the-world and Merleau-Ponty's (1968, 1945/2012) fleshy, reciprocal entwinements. In this post-Husserlian phenomenology, consciousness is superseded by a more inclusive, situated embodiment (micro-perception) in the midst of a culturally and historically colored world (macro-perception). Here, Ihde inserts technology into a revised intentionality equation: being-in-the-world or human-world becomes human-technology-world. Our relational yoke to the world is always already *mediated* or "technology-textured" (Ihde, 1990, p. 163). Or as Heidegger (1951/1971) put it, "we are the bethinged, the conditioned ones" (pp. 178–179).

Ultimately, Ihde (2008a) distinguishes his approach from classical phenomenology by undertaking empirical investigations of specific technologies:

> What these [investigations] taught me was how diverse they were, how differently embedded in different cultures even the same technologies may be, and above all, how technological history is so full of surprises, the unexpected and with unintended and unpredictable side-effects. (Ihde, 2008a, p. iv)

For Ihde, these in-depth phenomenologies of individual technologies mark phenomenology as *post*phenomenology. His project is also underwritten by several key insights about technology including:

1. *Multistability*: Any given technology may be used in multiple different ways across different contexts; as such, multistability undercuts technological determinism whether of the dystopian or utopian variety.
2. *Human-technology relations*: Our human-technology-world relations have several focal variations which Ihde has named embodiment, hermeneutic, and alterity, but also background relations. Verbeek (2008) has augmented these variations. In postphenomenology, the intentional relation is denoted by "human-technology-world" with varying sets of brackets and arrows; of particular interest is what happens in the interfacial hyphen or "enigma" position where (human-technology) or (technology-world) are intertwined most intimately.
3. *Amplification-reduction structure*: Every technology extends our perceptual and/or actional possibilities while simultaneously attenuating or atrophying others.
4. *Revealing-concealing structure*: Derived from Heidegger's ready-to-hand (handy but invisible)/present-at-hand (broken and obstinate) hammer analysis, our being-in-the-world of equipment is primarily one of concealment (Kiran & Verbeek, 2010, p. 417). Hammering, the hammer withdraws; that is, in use, the hammer is not the figure of our attention but is instead absorbed in the taken-for-granted background of our lifeworld. Thus to study a technology, one must be able to relinquish one's present-at-hand grip on the technology of interest in service of learning the pre-predicative language of its handiness in everyday life.

The following *methods* have also been explicitly associated with postphenomenology:

1. *Variational method or analysis*: Husserl's eidetic reduction or "imaginative or free variation" is adapted by Ihde, whereby the researcher may reveal a technology's multistabilities as opposed to Husserlian "essences" (Ihde, 2009; Rosenberger, 2014). Elsewhere, Rosenberger and Verbeek (2015) describe it as "the method of brainstorming stabilities of a multistable technology" (p. 27).
2. *Variational cross-examination*: Rosenberger (2014) suggests that a postphenomenologist must first determine an object's dominant stability and then perform a "critical contrast of particular stabilities that have been identified" to understand more about these alternatives (p. 382). This involves examining: 1) comportments and habits—the way a technology

relates to the body and a person's perception, 2) role within a program—the role of a particular technology within a network of actors, and 3) concrete tailoring—the physical alterations of technology for different purposes (pp. 382–385).

3. *Case study*: More broadly, Rosenberger and Verbeek (2015) characterize postphenomenology as a "case study approach" to research (p. 32). Case study has become a popular way to frame postphenomenological studies. See, for example, Irwin's (2016) multiple chapter "cases"—from earbud embodiment to self-tracking technologies.
4. *Conversational Analysis (CA):* Verbeek (2016) recently forwarded Conversational Analysis (CA) as "the missing link" for studying empirically not only "what things do," but how technologies mediate our perceptions and actions "from within" (p. 195). Stemming from ethnomethodology, CA is offered as a possible bridge method between postphenomenology and ANT.

In the next section, we look at postphenomenology in educational research, and then return to the question of method.

Postphenomenology, Technology, and Education: A Literature Review

Education's scholarly literature has long been aware of Ihde's phenomenology of technics. In 1985, for example, John W. Murphy cited Ihde (1979) and declared that, "without an examination of the technological 'world-view', technicized images of the classroom, learning, and students which accompany it may go undetected" (p. 167). That same year, John M. Broughton (1985) and Edmund V. Sullivan (1985) both made note of a lesser-known chapter of Ihde's, titled "A Phenomenology of Man-Machine Relations," that appeared in the edited volume *Work, Technology and Education* (1975). Sullivan observes that Ihde's work should lead educators to ponder how taken-for-granted scholastic objects like projectors, microscopes, and typewriters, "when . . . working properly . . . are on the periphery of your awareness, accomplish their function in a quiet and unobtrusive manner" (p. 1). Computing technology, it seemed, was poised to "be a figure as opposed to ground in the educational ecology" (p. 1). The work of other phenomenological philosophers such as Hubert Dreyfus, Heidegger, Merleau-Ponty also figured in these early articles; each intended to wake educators up to their technologized surround; each was taking a critical crack at Seymour Papert's (1980) *Mindstorms: Children, Computers and Powerful Ideas.* A decade later, Jeanne Connell (1996) devoted an entire article to the pedagogical implications of Ihde's philosophy of technology, and concluded that "educators must increase their awareness of the ways that computer technologies modify, extend, and transform the nature of experience in the classroom" (p. 12). However, none of these texts offered a sustained empirical study of individual technologies in education, only philosophical argument.

Empirical Studies

In the late 1980s, some phenomenologies of technics in education began to surface. One such example is Stefan Baldurssen's (1989) dissertation on the phenomenology of teaching and learning writing via word-processing software. Citing Ihde (1979, 1983), Baldurssen (1989) argues that assessments of technology in general are not adequate and that investigation of specific educational technolog*ies* and their mediations by way of phenomenology is critical. His empirical study of word-processing in schools, however, drew nothing from Ihde in terms of his methods of inquiry. Instead, as van Manen's doctoral student, Baldurssen adopted an early version of van Manen's practical phenomenology, which was published soon after as *Researching Lived Experience: Human Science for an Action Sensitive Pedagogy* (van Manen, 1990).

Since that time, van Manen has supervised multiple dissertations involving practical phenomenologies of technologies in education. One notable example is Norm Friesen's (2003) *Student-Computer Relation: A Phenomenology of its Pedagogical Significance*. A portion of his dissertation, a phenomenology of simulated frog dissection in the classroom, appeared in a 2011 issue of *Techne* (Friesen, 2011a). Friesen's work drew critique from Ihde (2012a) and Rosenberger (2012) who argued that computer simulations do not simply limit possibilities but may also provide richer and expanded possibilities. Friesen countered with a discussion of how technology can bring a hyperreal quality to instruction, whereby students may miss out on a more embodied experience and leave students less connected to the world around them (Friesen, 2011b).

Cathy Adams also studied with van Manen. Her pedagogy of PowerPoint revealed some of the ways in which student-teacher relations are mediated through this ubiquitous presentation technology (Adams, 2010). Teachers' practices may be supported and enriched by using PowerPoint in their classrooms, but they are also "enmeshed and relinquished to the language, imagery, framing, at-handedness and sensuality of their material and design" (Adams, 2010, p. 1). She also showed how PowerPoint may incline how a class unfolds. Other phenomenologies of specific technologies include van Manen and Adams's (2009) study of writing online, and van Manen's (2010) article on privacy in social media spaces or "Momus technologies" (p. 1023). Over the last few years, Adams has supervised several doctoral students claiming Ihde's and Verbeek's postphenomenology as their theoretical framework or primary methodology. Some have struggled to articulate its specifics beyond van Manen's phenomenology of practice, or too, in relation to some of the theoretical concepts and methods associated with ANT, such as "generalized symmetry" and "following the actors."

Postphenomenology and Education

A few educational researchers have also begun to publish their educational inquiries explicitly as *post*phenomenology. These scholars include Jesper Aagaard (2015a, 2015b, 2016), Catherine Hasse (2008, 2015a, 2015b), Peggy Jubien (2014, 2015), and Tobias Röhl (2012a, 2012b, 2015). Aagaard's (2015a) postphenomenological research concentrated on upper secondary students' use of technology in college classrooms. He found that use of digital devices is so ubiquitous that many students do not bring books or even pens to school. After several months of observing students using their laptops for off-task activities during lectures, he began formal interviews with students. His interview data revealed several themes: (1) technology is a mixed blessing—its affordances include speed of communication and access to information but the same technology provides a "backdoor" where students can escape from class without leaving the room (Aagaard, 2015a, p. 93); (2) students are drawn to distraction—devices provide an allure that students describe as hard to resist; (3) off-task technology use tended to occur when a lesson was considered too difficult or boring, or when there was a break (for example, waiting for the professor to set something up) (pp. 93–94). He concludes that "we need a deliberate shaping of our involvements with technologies to develop a free relation to them" (p. 96). He suggests that this "free relation" may be supported by implementing strategies such as taking regular technology breaks during class or prompts from the teacher to "close the lid." Such studies are crucial in understanding the meaning of quantitative studies, for example, Carter, Greenberg and Walker's (2016) recent finding that everyday device use in postsecondary classes lowers exam results. While the Carter et al.'s study examined students' performance on standardized exams in classrooms that permitted the use of digital devices to take notes compared to those that did not, it was silent on students' everyday experiences of using technology in the classroom.

In "Methods of Materiality: Participant Observation and Qualitative Research in Psychology," Aagaard and Matthiesen (2016) challenge the privileging of "human attribution of meaning" in qualitative research (p. 34). They call for a turn to the material world in order to account for the reciprocal intertwinings of meaning and materiality. Artifacts are not neutral, unsubstantial things that simply receive whatever meanings we attribute to them, but rather material presences that also act upon us:

> To be clear, we are not trying to throw out the qualitative baby with the bathwater, and this third space is decidedly not some sort of demilitarized zone between interpretive and causal approaches. We explicitly situate material presence within the realm of qualitative inquiry. . . . Our argument is not *against* interpretation but *for* an oscillation between linguistic meaning and material presence. (Aagaard & Matthiesen, 2016, p. 36)

Aagaard and Matthiesen (2016) point to an empirical example of authority in the classroom during parent-teacher interviews with Somali parents in public schools in Denmark, to illustrate how the materiality of the classroom perpetuates a particular social order and influences action. They argue that research that employs interviews "without capturing our everyday intertwinement with material artifacts" (p. 40) may render an incomplete picture. To address this lack of attention to materiality, qualitative researchers must develop both "an *ear for meaning* and an *eye for materiality*" (p. 41). They outline three possible ways to accomplish this:

1. use posthumanist analytic strategies,
2. study material presence in terms of agency/structure, and
3. draw situational maps.

(Aagaard & Matthiesen, 2016, p. 41)

For the first, they point to Röhl's (2015a) work. Röhl (2012a), in an ethnography of the materiality of two secondary schools in Germany, explores the ways in which people and objects are interwoven and impact one another. He identifies a gap in understanding materiality, where studying objects has focused largely on symbolic meaning, rather than understanding "what things do" (Verbeek, 2005), in other words, a postphenomenological approach. He proposes three conceptual strategies for examining materiality in classrooms using a combination of ANT, SCOT, and postphenomenology: respectively symmetry, emergence, and "broken hammer" or breakdown. Röhl also invokes postphenomenology where relations between humans, objects, and the world are examined. His primary concern is with how objects are involved in knowledge creation, and how certain materials "enable and constrain certain forms of knowledge" (Röhl, 2012a, p. 122). To this end, he proposes a "material ethnography" and suggests postphenomenology may play a role in the investigation of "the embodied and sensory relation that is constituted between human actor and material object" (pp. 121–122).

In her "Postphenomenology: Learning Cultural Perception in Science," Hasse's (2008) interest is in revealing the embodied and situated knowledge practices of physicists. She frames her study in terms of Ihde's (2002) *body one* and *body two*, or the "lived sensuous and the cultured body" (Hasse, 2008, p. 45). Hasse says her quasi-experiments do not represent what might be considered typical phenomenological methods, but focuses on variational descriptions. "It is from these descriptions that [Hasse] unfolds an argument on the interaction of body one and body two and connects this perspective to learning in science education as well as scientific practice" (p. 50). She calls on Merleau-Ponty (1945/2002) and new material feminist scholarship to develop a vocabulary to assist in performing her analysis.

Jubien (2014) conducted two studies of smartphone usage in postsecondary education: (1) a phenomenological exploration of the experience of listening to pre-recorded lectures via podcasting; and (2) an ANT-based study of smartphone use in education. Similar to Rosenberger, she posited that phenomenology provides a way to collect the lived experiences of technology use, while ANT provides a complementary way to describe technological objects and their networked relations. Relying on Verbeek's postphenomenological vocabulary and also that of Ingrid Richardson's (2007) phenomenology of handheld devices or "pocket technospaces" (p. 205), Jubien conducted a third study under the rubric of postphenomenology. For Jubien, this ANT-phenomenology amalgam allowed for descriptions and analyses of how people and things may "join together, stabilize, work together, and break apart" (Jubien, 2014, p. 8). Nonetheless, her study revealed an unresolved or at least uneasy tension between the action-focused, material-semiotics of ANT, and the more existential, meaning saturated concerns of phenomenology.

To date, postphenomenology as a methodology in educational research is multistable. *Doing* postphenomenology in education has inspired an eclectic mix of human and social science methods, and is most typically evoked for its unique conceptual offerings and vocabulary regarding human-technology relations. Below we offer a *postphenomenology of practice*. We believe that this approach to inquiry may help make sense of some of the current empirical phenomenological studies of technology in education, as well as provide a well-articulated methodological backdrop for conducting postphenomenological research in education.

TOWARD A *POST*PHENOMENOLOGY OF PRACTICE

The postphenomenological research approach we outline here is strictly informed by van Manen's phenomenology of practice. The ambition of phenomenology of practice is simple: to describe and reflect on a phenomenon of professional or personal interest by attending to the prereflective or everyday lifeworld. Van Manen's (1990) *Researching Lived Experience,* along with early articles, sketched a way of doing educational research that diverged radically from the quantitative approaches of the day by advocating a "return to the [pedagogical] things themselves." Today, qualitative research is mainstream; and too, many now claim to be doing phenomenology in the human and social sciences. Van Manen's more recent (2014) *Phenomenology of Practice: Meaning-Giving Methods in Phenomenological Research and Writing*, published soon after he retired, is based on his years of teaching this form of human science inquiry to graduate students across multiple professional disciplines, and clarifies many methodological details of his approach.

Perhaps because phenomenology of practice is already pragmatic and empirical, and situated in the prereflective, concrete everydayness of things, researchers approaching technology via van Manen's phenomenology of practice had little need to lay it aside in favor of a *post*phenomenology. Nonetheless, recognizing that postphenomenology specifically signals "a nuanced change from classical phenomenology [by] thematizing of materiality, particularly in the form of instruments and devices by which we make 'worlds' available to us which were previously unexperienced and unperceived" (Ihde, 2003, p. 20), we easily adopt this term here. Phenomenology of practice, similar to Verbeek's (2015) methodological ambitions for postphenomenology, "combines philosophical analysis with empirical investigation" (p. 190).

In outlining a postphenomenology of practice, we begin by making a distinction between the *prereflective* (the natural attitude) and the *reflective* (the phenomenological attitude), which roughly corresponds to the familiar division employed in empirical research between data collection and data analysis. In a postphenomenology of practice, generating prereflective data consists of gathering lived experience descriptions (e.g., by way of interviews, journals, observation, etc.) and crafting *anecdotes* that describe a particular technology or thing as occurrent in everyday life. Reflective analysis proceeds by way of both human science methods (e.g., line-by-line, thematic, and existential analysis) and phenomenological methods (e.g., the epoché and the reduction) (see van Manen, 2014). A finished (post)phenomenological research text is a weave of both prereflective material and reflective insights intending to (1) emulate human-technology-world entwinements through textual description, and (2) explore and shed light on particular technology-texturings of pedagogical lifeworlds—relational, corporeal, spatial, temporal, material, and medial.

From a (post)phenomenological of practice perspective, the things of our immediate world speak to us prereflectively. The chair invites me to sit down, my e-mail tugs at me to check it, my buzzing iPhone insists that I answer it. Of course I may choose not to do any of these things. But having responded to the "call" of a thing, I am swiftly caught up in the particular world that the particular thing or technology opens for me: sitting, reading an e-mail, or talking to a friend on the phone. Our embodied selves always already dwell in the midst of this primordial "rapport with things" (Heidegger, 1951/1971, p. 157). These silent, ongoing corporeal conversations with our socio-material surround may be glimpsed by attending to the vocative or invitational quality of a given technology (Adams & Thompson, 2017). To access these human-technology-world correspondences requires attention to the prereflective or pathic dimensions of the lifeworld. Here, van Manen's (2014) phenomenology of practice provides multiple methods, derived from the human and social sciences, for gathering humans' prereflective rapport with their technologies by way of lived experience descriptions or anecdotes.

The Prereflective: Composing Anecdotes through Self-Observation (Phenomenological Protocol Writing)

A first way to generate data for a postphenomenological investigation is to attend to one's own lifeworld: "Our personal life experiences are immediately accessible to us in a way that no one else's are" (van Manen, 2014, p. 313). With practice, a phenomenological researcher begins to develop an eye for the non-intrusive observation of oneself and of others—so as not to disturb the moment or event in its unfolding "in the wild"—and too, an ear for the invitational appeals of one's equipmental or technology-textured surround. But to translate these prereflective moments into text (i.e., to *re-cord* with fidelity human-technology-world interactive/interpassive relations), demands a writerly attention to what appears in its prereflective lived-throughness. Explanations, opinions, judgments, or theoretical concepts must be pushed aside in favor of what was given in the moment. Such experiential happenings are necessarily written in retrospect, with attention to the sensuousness, situated detail of living through the moment.

Such concrete, lived-through descriptions are a crucial ingredient to a phenomenology, and more particularly postphenomenology. Indeed, Ihde's many postphenomenologies of technics are filled with such self-observant phenomenological description. For example, here Ihde (1979) describes the simple act of writing on a blackboard:

> I pick up . . . a piece of chalk and begin to trace it across the . . . blackboard. Upon a careful examination of this experience I suddenly discover that I experience the roughness of the board *at the end of the chalk*. This is, of course, also Merleau-Ponty's blind man who experiences the "world" at the end of his cane. If I begin to be descriptively rigorous, I find I must say that what I feel is felt locally at the end of the chalk, or better, at the chalk-blackboard junction. The

> "terminus" of my intentional extension into the world is on the blackboard, and I have discovered (contrary to empiricism) that touch is also a distant sense. (Ihde, 1979, p. 7)

Ihde opens with an otherwise unremarkable but recognizable "lived moment" to teachers: "I pick up . . . a piece of chalk and begin to trace it across the . . . blackboard" (1979, p. 7). Reflecting back with his postphenomenologist's eye, that is, momentarily loosening "the intentional threads that connect us to the world in order to make them appear" (Merleau-Ponty, 1945/2012, p. xxvii), Ihde notices that in performing this taken-for-granted action, "I experience the roughness of the board *at the end of the chalk*" (1979, p. 7). He then reflects further on this briefest of moments in more phenomenological depth, drawing in more precise and concrete prereflective material ("what I feel is felt locally at the end of the chalk, or better, at the chalk-blackboard junction" [ibid]) with the assistance of Merleau-Ponty and his then nascent postphenomenological vocabulary.

In a recent study examining writing technologies in schools, Adams (2016) also re-assembles a lived-through moment based on her self-observation:

> My hand, or rather hand*s* at the desktop computer (or laptop or iPad) keyboard *want* to write, and too, they want to write together. Rhythmic taps and clustered bursts of understanding live effortlessly between them. They have long since established a unique corpus of keystroke dances and jigs, as singular as my handwritten signature. Together my two hands' fingers patter out letters, words, and sentences, dividing their choreographic work seamlessly among themselves. The right hand, in command of the drop-down menus, tabs, toolbar and scroll, moves fluidly back and forth between mouse and keyboard. There is no question of encroaching on one another's space: my fingers and hands *are* the space. (Adams, 2016, p. 483)

Such first-person description is developed by paying close attention to the *language of keyboarding* in order to re-evoke some of its sensual and gestural aspects (e.g., the soft "patter" generated by the fingers moving swiftly from key to key, and the "clustered bursts" of a writer in the throes of composition). Attending to the vocative dimensions of textual description is also a key dimension of a (post)phenomenology of practice (van Manen, 2014, pp. 240–296).

The Prereflective: Gathering Lived Experience Descriptions through Interviews

Another way to gain access to the lived world of others is to conduct interviews with those who have experience with the technology-textured phenomenon in question. Interviewing human participants is one of the most widespread methods of data generation in qualitative research practice today. However not all research interviews are alike. In a postphenomenology of practice interview, the primary purpose is to elicit lived experience descriptions (LEDs) about the research participant's everyday engagements and encounters with the technology of interest. As such, the researcher asks questions to assist the participant in recalling specific moments, and then in providing a lived-through account of the event. Interviewing for LEDs requires a particular kind of questioning. Investigating the experience of teaching with PowerPoint (Adams, 2010), for example, the researcher might offer the following prompts: "Can you think back to the last time you taught with PowerPoint? . . . Recalling that particular class, can you walk me through what happened, perhaps beginning with when you arrived and opened your Powerpoint?" Or, "Do you have a particular moment that stands out when you were teaching with PowerPoint? Where were you? What happened?" Interviewees may also be invited to write down a specific recollection in lived-through detail. Interview transcriptions are subsequently culled for LEDs.

Aagaard (2015a) employs interviews in his postphenomenological studies. He also worries that, "despite an increased sensitivity to embodied use of technologies, postphenomenologically oriented scholars rarely conduct empirical studies of other people's technologically mediated experiences and practices" (p. 92). Interviewing, sometimes combined with observation of others engaged in practice, can be an excellent way to generate multiple experiential variations or multistablities in relation to a particular technology, as well as to potentially disrupt one's certainty about how a technology may or may not appear or be used. For a postphenomenology of practice, interviewing others is a core method for generating a rich and diverse body of first-hand examples of everyday human-technology-world engagements (see also Adams & Thompson, 2017, pp. 24–33, on "gathering anecdotes").

The Prereflective: Composing Anecdotes through Observation of Others

Yet another way to generate postphenomenological data is through observing others in the context of their technologized surround. This data generation strategy, popular in ANT as well new materialist and nonrepresentationalist ethnographies, is also favored by some postphenomenologists in education. While observation of others does not give access to another's prereflective experience as such, it may assist in pointing up aspects of everyday life that may otherwise be taken for granted by oneself and others. Aagaard (2015b) employed this approach by first observing device use in classrooms, and later interviewing students. Observing others engaged in their equipmental surround can be invaluable in spotting what may otherwise be invisible in one's own technology practices, and thus allows the researcher to lift gently to notice what would otherwise pass without remark (i.e., by slackening "the intentional threads which connect us to the world in order to make them appear" [Merleau-Ponty, 1945/2012, p. xxvi]).

We suggest, however, that observation of others or even of self may not always constitute sufficient data to (re)construct certain phenomenologies of technology. In his ethnography of a science classroom, Röhl (2012b) relies on his disciplined observational skills to write rich and detailed accounts of events. He then evokes postphenomenology to address the corporeal and sensorial world of the students:

> By observing students and teachers and how they engage material objects in the classroom, we can try to describe and analyse how the objects' sensory and bodily dimension affects them. . . . Postphenomenological analyses can also be understood as descriptions of how the researcher relates to the material objects. The researcher partly assumes the role of a participant in science classes. (Röhl, 2012b, p. 54)

Here Röhl continues his ethnographic stance, and attempts to situate himself so he may speak on behalf of the students and thereby describe the invitational quality and prereflective dimensions of their lifeworld:

> When I visited the science classes, I usually sat on a chair at one of the desks, listened to classroom discourse, observed experiments and copied what was written on the blackboard into my notebook—much like the students I observed. Thus, I was similarly affected by the material objects used in the classroom and can use that perspective to describe how demonstration experiments and the blackboard invite us to use them in a specific way. (Röhl, 2012b, p. 54)

Yet, it is ultimately not the researcher, but the students, who are best positioned to describe how the world of the classroom and the science experiment appeared to them, that is, the intelligibility of the event to each individual, what material aspects of the experiment showed up as meaningful to them, and so on. Here, a postphenomenology of practice necessarily returns, not only to the things themselves, but also to the students themselves. While the observant researcher can indeed glean some aspects of another's technology-textured lifeworld by carefully reporting their own perceptual apprehensions of a situation, the postphenomenologist is charged with pursuing a technology's multistabilities. To accomplish this, observation must also be combined with other methods in order to reveal the unique human-technology-world that is opened by a given technology for a given individual in a given context.

In the final section, we examine how one may *begin* to *reflect* on these gathered prereflective experiences in a postphenomenology of practice. In qualitative research, such phenomenological reflection is roughly equivalent to data analysis or interpretation. For Merleau-Ponty (1945/2002, p. xvii), the ambition of phenomenological reflection is to "emulate the unreflective life of consciousness":

> Reflection does not withdraw from the world toward the unity of consciousness as the world's basis; it steps back to watch the forms of transcendence fly up like sparks from a fire; it slackens the intentional threads which attach us to the world and thus brings them to notice; it alone is consciousness of the world because it reveals that world as strange and paradoxical. (Merleau-Ponty, 1945/2012, p. xv)

In phenomenology, such reflection is achieved by way of the reduction, or more precisely, the *epoché-reduction* couplet. The epoché-reduction is a twofold methodological gesture that intends to at once suspend one's preconceptions (i.e., the epoché) in order to discover the experiential surge of the lifeworld (i.e., the reduction proper). Here it is crucial that the reader divest themselves of the erroneous belief that the phenomenological reduction is about reducing a phenomenon down to some single essence, or as Aagaard (2015a) describes, "boil[ing it] down to a stock cube of *essential meaning*" (p. 92, italics in original). In line with the etymological roots of the term "reduction" (to bring back or restore), the phenomenological reduction also aims to restore or at least resemble the originary phenomenon in its lived wholeness. This sometimes surprising meaning of the reduction is also evident in medical terminology. Reduction is the name of the surgical procedure for resetting a broken bone, that is, it is the process of restoring a bone to its original form, placement, and functioning. In phenomenology, the reduction similarly leads us back to life in its prereflective wholeness, as it was originarily lived. We will return to this notion of the reduction as restoration or "reassembled resembling" (Adams & Thompson, 2017, p. 31) shortly. For now we note that this type of research proceeds not by deduction or induction but by reduction (van Manen, 2014, p. 218).

In his *Experimental Phenomenology*, Ihde (2012b) similarly engages the epoché-reduction couplet, and describes it by a simpler term: "phenomenological looking."

> The first steps of phenomenological looking are usually called an *epoché*, which means to suspend or step back from our ordinary ways of looking, to set aside our usual assumptions regarding things. Within this general stance, particular levels of stepping back are then determined; these levels are termed *phenomenological*

> *reductions*. I shall interpret these specifications as working rules or directions for the way the investigation may proceed. Thus, *epoché* and phenomenological reductions may also be called hermeneutic rules, since they provide the shape or focus of the inquiry. Hermeneutic in its broadest sense means interpretation, and rules give shape to an interpretation. (Ihde, 2012b, p. 17)

In his phenomenology of practice, van Manen (2014) outlines human science methods (e.g., examining an LED line-by-line, thematically, and existentially) as well as philosophical methods (e.g., performing multiple epoché-reductions) to assist with "phenomenological looking" or reflecting on the prereflective. The *eidetic* reduction, for example, recognizes that every phenomenon obtains its unique identity via its differences from other related phenomena. A phenomenon is originally constituted and marked by what it is *not*. The eidetic reduction is sometimes performed by varying the example given by an LED. The intent is to uncover aspects of a phenomenon that are *invariant* or unique. Ihde makes extensive use of this "varying the example" technique in his many fine analyses of technologies in use. Readers familiar with the vocabulary of postphenomenology will recognize the eidetic reduction as similar to variational method or analysis used to uncover the multistabilities of a given technology (Ihde, 2009; Rosenberger, 2014; Verbeek, 2005).

Below we address another example of reflective analysis via the epoché-reduction, by using a technique found across empirical studies in the human and social sciences including phenomenology and postphenomenology, but also ANT (e.g., Latour's [1996] *Aramis* project), ethnomethodology (e.g., Harold Garfinkel's [1967] breaching experiments), and media ecology (McLuhan, Hutcheon & McLuhan's [1977] *City as Classroom* projects): studying breakdowns. Paying close attention to what transpires when a technology breaks down, malfunctions, or is missing is also a reliable way to perform the eidetic reduction.

The Reflective: Studying Breakdowns and the Eidetic Reduction

Sometimes, like David Mitchell (played by Jake Gyllenhaal) in the film *Demolition* (Sipe & Vallée, 2015), we need a broken hammer to dislodge us from our taken-for-granted lodge in life, and set us on a new path. Only in this way may we hope to encounter what is otherwise hidden from us, or of what we possess only a distant awareness. Suffering a psychological collapse after the war, Heidegger wrote in his 1946 "Letter on 'Humanism'": "'Philosophizing' about a breakdown [*das Scheitern*] is separated by a chasm from a broken down thinking. If this should fortunately come to a person, what would occur is no misfortune. To him would come the sole gift that could come to thinking from being" (Heidegger in Mitchell, 2015, p. 21).

As Andrew Mitchell paraphrases, "Thinking must be broken open . . . the gift of the break down is a break through into the world" (2015, p. 21). Such is the gift of Heidegger's (1962) broken hammer analysis in granting us primordial entrance to what is most near: "things" and to "the being of the entities encountered in the environment" (p. 95). The event of a breakdown—whether a tool, one's body, or a relation to another—may momentarily wrest from us the stubbornness of our preconceptions (the epoché) and open the possibility to recuperate the lifeworld as it is meaningfully constituted (the reduction). A breakdown is thus a kind of naturally occurring epoché-reduction: what is taken for granted falls temporarily away, and meaning structures that were in play briefly collapse. For a moment, what we take most for granted may be made potently visible through absence. In the context of ANT, for example, networked assemblages and unexpected alliances may be revealed when the binding ties are suddenly severed. In the context of postphenomenology, a breakdown provides a reliable way of surfacing our otherwise taken-for-granted, co-constitutive relations with technologies and thus make them available for inquiry (Adams & Thompson, 2017).

Studying breakdowns or employing the "broken hammer" strategy is noted by Röhl (2012a) as well as Aagaard and Matthiesen (2016). For Röhl, "the 'broken hammer' reveals the manifold relations educational 'tools' are part of and the way objects are treated in order to overcome their material resistance" (Röhl, 2012a, p. 119). He provides two examples. The first is a science experiment that did not go as planned, the second involves a teacher who uses an object for an unexpected purpose. In both instances the "broken hammer" is employed by Röhl as a *concept* to help explain the fragility of epistemic or knowledge objects in classrooms, to identify material "resistances" and their visibilities (present-at-hand), and how these resistances may be overcome (i.e., returned to ready-to-hand) for pedagogical effect. However, this approach to studying breakdown is not what is intended by the eidetic reduction; rather it demonstrates a way to leverage the insights of (post)phenomenology in order to perform a sociomaterialist analysis.

By way of comparison, in her analysis of scholastic writing instruments, Adams (2016) employs the occasion of her broken hand to perform an eidetic reduction. She varies the example of writing with her "non-writing" hand as well as with different instruments. In the process, she reveals how a writing *templum* is convened by the knowing, writing but prereflective body in congress with its instruments. Citing Merleau-Ponty's (1945/2012) example of an experienced organist adjusting to a new organ, Adams writes: A similar gestural rehearsal, resizing, and resettling unfolded as I tried keyboarding with my injured hand. My writing instruments—keyboard, mouse, screen, and word processor—had not changed, but one of my hands had. As a "subject who [had already learned] to type," I had "literally incorporate[d] the space of the keyboard into [my] bodily space" (Merleau-Ponty, 1945/2012, p. 146). Trying to keyboard and mouse in this somewhat modified material landscape, "new knot[s] of significations" began to form as my writerly body adjusted, returned, and reorganized. With a little practice, my new hand and finger situation were quickly inducted into and absorbed in the new arrangement. An adjusted, but nonetheless wholly familiar, writing *templum* once again convened. (Adams, 2016, p. 484)

Having revealed writing—no matter its instruments—as the convening of a particular kind of practice through varying the example via a broken-handed writing body, Adams now turns to the nuanced atmospheres opened by different writing instruments in the midst of writing. As Ash and Simpson (2014) maintain, a sustained attention to our lived corporeal engagements with things is a crucial aspect of postphenomenological research. Postphenomenology requires attention to embodied experience *alongside* "an emphasis on the ways in which the body-subject undergoes constant processes of 'affectual composition' in and through its relations with a material-agential world" (Ash & Simpson, 2014, p. 8). Revealing the dynamism of the lived body's everyday engagements with its instruments demands a return to phenomenological methods (such as the eidetic reduction) in order for "unseen things [to] become 'visible'" (Ihde, 2003, p. 20), and for the uniqueness of the "revealing-concealing structure" (Kiran & Verbeek, 2010, p. 417) of a particular technology in relation to its human user to be made evident.

CONCLUSION

In educational technology circles, the mantra, "it's just a tool" is still used to describe the belief that technology is neutral and that the teacher or student alone is responsible for constructing educational meaning. Postphenomenology provides a welcome ameliorative to this sovereign humanist position, foregrounding our co-constitutive relations with our technological surround. Gathering and reflecting on phenomenological descriptions of using technology in pedagogical situations can provide much needed insight into the mediating influences of the wide range of technologies. As such, postphenomenological studies and insight may serve in promoting more critically circumspect applications of different technologies in pedagogical settings, and in advancing a long-overdue revision to our taken-for-granted assumptions and practices with technologies in education.

Doing postphenomenogy in education involves attending to the unique differences a particular technology makes to teaching practice, knowledge apprehension, and pedagogical meaning. To date, postphenomenological methods have included data collection techniques such as ethnographic observation, self-observation, and semi-structured interviewing, and analytic techniques such as the "broken hammer" and variational theory. Adopting a phenomenology of practice (van Manen, 2014) provides another empirical, qualitative approach to examine carefully and critically the mediating effects of individual technologies. A *post*phenomenology of practice recovers pre-reflective experiences of our everyday relations with technology as crucial data in its empirical studies. It also reinstates the epoché-reduction couplet as a core philosophic method in postphenomenology's repertoire of heuristics. The place of the "broken hammer" is thus clarified not only as a concept for interpreting socio-material results, but also as a way of proceeding reflectively in performing variational analysis.

In their "Field Guide to Postphenomenology," Rosenberger and Verbeek (2015) tell us that postphenomenology follows no strict method. Nonetheless, they have observed that postphenomenological studies seem to share some characteristics such as including "empirical work as a basis for philosophic reflection" (p. 31) and a focus on how technologies mediate human-world relations. For Rosenberger and Verbeek, by "applying postphenomenological concepts to concrete instances of human-technology relations, the advantages, disadvantages, limits, and places of potential expansion and enrichment can be identified" (p. 32). We suggest that van Manen's (2014) "phenomenology of practice" provides another approach to postphenomenology that leans heavily on its phenomenological roots while providing a stable methodological ground for conducting empirical postphenomenological studies in education.

REFERENCES

Aagaard, J. (2015a). Drawn to distraction: A qualitative study of off-task use of educational technology. *Computers & Education*, *87*, 909–907. doi:10.1016/j.compedu.2015.03.010

———. (2015b). Media multitasking, attention, and distraction: A critical discussion. *Phenomenology and the Cognitive Sciences*, *14*(4), 885–896. doi:10.1007/s11097-014-9375-x

———. (2016). Breaking down barriers: The ambivalent nature of technologies in the classroom. *New Media and Society*, 1–17. doi 10.1177/1461444816631505

Aagaard, J., & Matthiesen, N. (2016). Methods of materiality: Participant observation and qualitative research in psychology. *Qualitative Research in Psychology*, *13*(1), 33–46. doi:10.1080/14780887.2015.1090510

Adams, C. (2010). Teachers building dwelling thinking with slideware. *Indo-Pacific Journal of Phenomenology*, *10*(1), 1–12. doi:10.2989/IPJP.2010.10.1.3.1075

———. (2016). Programming the gesture of writing: On the algorithmic paratexts of the digital. *Educational Theory*, *66*(4), 479–497. doi:10.1111/edth.12184

Adams, C., & Pente, P. (2011). Teachers teaching in the new mediascape: Digital immigrants or "natural born cyborgs"? *E-Learning and Digital Media*, *8*(3), 247–257.

Adams, C., & Thompson, T. L. (2017). *Researching a Posthuman World: Interviews with Digital Objects*. Basingstoke and London, England: Palgrave MacMillan.

Adams, C., & van Manen, M. (2008). Phenomenology. In L. Givens (Ed.), *The SAGE Encyclopedia of Qualitative Research Methods* (pp. 614–619). Thousand Oaks, CA: SAGE Publications.

Ash, J., & Simpson, P. (2014). Geography and post-phenomenology. *Progress in Human Geography*, *38*(1), 1–19. doi:10.1177/0309132514544806

Baldursson, S. (1989). *Technology, Computer Use, and the Pedagogy of Writing* (Doctoral Dissertation). Available from ProQuest Dissertations & Theses Global (Order No. NL52929)

Broughton, J. M. (1985). The surrender of control: Computer literacy as political socialization of the child. In D. Sloan (Ed.), *The Computer in Education: A Critical Perspective* (pp. 102–122). New York, NY: Teachers College Press.

Carter, S. P., Greenberg, K., & Walker, M. (2016, May). *The Impact of Computer Usage on Academic Performance: Evidence from a Randomized Trial at the United States Military Academy. SEII Discussion Paper #2016.02*. Cambridge, MA: MIT Department of Economics & National Bureau of Economic Research. Retrieved from https://seii.mit.edu/wp-content/uploads/2016/05/SEII-Discussion-Paper-2016.02-Payne-Carter-Greenberg-and-Walker-2.pdf

Connell, J. M. (1996). Exploring some of the educational implications of Ihde's philosophy of technology. *Journal of Educational Foundations, 10*(4), 5–12.

Friesen, N. (2003). *Student-Computer Relation: A Phenomenology of its Pedagogical Significance* (Doctoral Dissertation). Available from ProQuest Dissertations & Theses Global (Order No. NQ87972)

———. (2011a). Dissection and simulation: Brilliance and transparency, or encumbrance and disruption? *Techne: Research In Philosophy & Technology, 15*(3), 185–200.

———. (2011b). Less is more: A response to Ihde, Rosenberger, Borgmann, Barney and Sørensen. *Techne: Research In Philosophy & Technology*, *15*(3), 229–234.

Garfinkel, H. (1967). Studies of the routine grounds of everyday activities. In *Studies in Ethnomethodology* (pp. 35–75). Cambridge, MA: Polity Press.

Hasse, C. (2008). Postphenomenology: Learning cultural perception in science. *Human Studies, 31*(1), 43–61. doi:10.1007/s10746-007-9075-4

———. (2015a). *An Anthropology of Learning: On Nested Frictions in Cultural Ecologies*. New York, NY: Springer.

———. (2015b). The material co-construction of hard science fiction and physics. *Cultural Studies of Science Education, 10*(4), 921–940.

Heidegger, M. (1962). The worldhood of the world. In *Being and Time* (J. Macquarrie & E. Robinson, Trans.), 63–148. New York, NY: Harper & Row.

———. (1971). *Poetry, Language, and Thought* (A. Hofstadter, Trans.). New York, NY: Harper & Row. (Original work published 1951)

Ihde, D. (1975). A phenomenology of man-machine relations. In W. Feinberg & H. Rosemont (Eds.), *Work, Technology and Education* (pp. 186–203). Urbana, IL: University of Illinois Press.

———. (1979). *Technics and Praxis: A Philosophy of Technology*. Dordrecht: Reidel.

———. (1983). *Existential Technics*. Albany, NY: State University of New York Press.

———. (1990). *Technology and the Lifeworld: From Garden to Earth*. Bloomington, IA: University Press.

———. (2002). *Bodies in Technology*. Minneapolis, MN: University of Minnesota Press.

———. (2003). *Postphenomenology—Again?* Working Papers, Aarhus University, Denmark, pp. 3–25. Retrieved from http://sts.au.dk/fileadmin/sts/publications/working_papers/Ihde_-_Postphenomenology_Again.pdf

———. (2006). Technofantasies and embodiment. *Critical Studies*, *29(*1), 53–166.

———. (2008a). *Ironic Technics*. Copenhagen, Denmark: Automatic Press/VIP.

———. (2008b). Aging: I don't want to be a cyborg! *Phenomenology and the Cognitive Sciences*, *7*(3), 397–404.

———. (2009). *Postphenomenology and Technoscience: The Peking University Lectures*. Albany, NY: Suny Press.

———. (2012a). Dissection and simulation: A postphenomenological critique. *Techné: Research in Philosophy & Technology, 15*(3), 203–205.

———. (2012b). *Experimental Phenomenology: Multistabilities* (2nd ed.). Albany, NY: State University of New York.

———. (2016). From phenomenology to postphenomenology. In *Husserl's Missing Technologies* (pp. 122–134). New York, NY: Fordham University Press.

Irwin, S. O. (2016). *Digital Media: Human-Technology Connection*. Lanham, MD: Lexington.

Jubien, P. (2014). Postphenomenology, smartphones, and learning: Students and teachers in higher education. *SAGE Research Methods Cases*. London, England: SAGE Publications. doi:10.4135/978144627305013512947

———. (2015). *A Postphenomenological Exploration of Mobility in Post-secondary Teaching and Learning*. Doctoral Dissertation. Edmonton: University of Alberta.

Kiran, A. H., & Verbeek, P.-P. (2010). Trusting our selves to technology. *Knowledge, Technology & Policy, 23*(3/4), 409–427. doi:10.1007/s121300–109–123-7

Latour, B. (1992). Where are the missing masses? A sociology of a few mundane artifacts. In W. E. Bijker & J. Law (Eds.), *Shaping Technology/Building Society: Studies in Sociotechnical Change* (pp. 225–258). Cambridge, MA: The MIT Press.

———. (1996). *Aramis, or the Love of Technology*. (C. Porter, Trans.) Cambridge, MA: Harvard University Press.

McLuhan, M., Hutchon, K., & McLuhan, E. (1977). *City as Classroom: Understanding Language and Media*. Agincourt, Ontario: Book Society of Canada.

Merleau-Ponty, M. (1968). *The Visible and the Invisible* (A. Lingis, Trans.). Evanston: Northwestern University Press.

———. (2002). *Phenomenology of Perception* (C. Smith, Trans.). New York, NY: Routledge. (Original work published 1945)

———. (2012). *Phenomenology of Perception* (D. A. Landes, Trans.). New York, NY: Routledge. (Original work published 1945)

Mitchell, A. J. (2015). *The Fourfold: Reading the Late Heidegger*. Evanston, IL: Northwestern University Press.

Murphy, J. W. (1985). Consideration of computer mediated education: A critique. *Phenomenology + Pedagogy, 3*(3), 167–176. Available: https://ejournals.library.ualberta.ca/index.php/pandp/article/view/14988/11809

Papert, S. (1980). *Mindstorms: Children, Computers, and Powerful Ideas*. New York, NY: Basic Books.

Richardson, I. (2007). Pocket technospaces: The bodily incorporation of mobile media. *Continuum: Journal of Media and Cultural Studies, 21*(2), 205–215. doi:10.1080/10304310701269057

Röhl, T. (2012a). Disassembling the classroom—An ethnographic approach to the materiality of education. *Ethnography & Education, 7*(1), 109–126. doi:10.1080/17457823.2012.661591

———. (2012b). From witnessing to recording—Material objects and the epistemic configuration of science classes. *Pedagogy, Culture & Society, 20*(1), 49–70.

———. (2015). Transsituating education: Educational artefacts in the classroom and beyond. In S. Bollig, M.-S. Honig, S. Neumann, & C. Seele (Eds.), *MultiPluriTrans in Educational Ethnography: Approaching the Multimodality, Plurality and Translocality of Educational Realities* (pp. 143–161). Beilefeld, Germany: Transcript.

Rosenberger, R. (2012). A phenomenological defense of computer-simulated frog dissection. *Techné: Research in Philosophy & Technology, 15*(3), 209–221.

———. (2014). Multistability and the agency of mundane artifacts: From speed bumps to subway benches. *Human Studies, 37*(3), 369–392. doi:10.1007/s10746-014-9317-1

Rosenberger, R., & Verbeek, P.-P. (2015). A field guide to postphenomenology. In R. Rosenberger & P.-P. Verbeek (Eds.), *Postphenomenological Investigations: Essays on Human–Technology Relations* (pp. 9–41). Lanham, Maryland: Lexington Books.

Sipe, B. (Writer), & Vallée, J.-M. (Director). (2015). *Demolition* [motion picture]. United States: Fox Searchlight Pictures.

Sullivan, E. V. (1985). Computers, culture, and educational futures—A meditation on *Mindstorms*. *Interchange, 16*(3), 1–18.

van Manen, M. (1990). *Researching Lived Experience: Human Science for an Action Sensitive Pedagogy*. Albany, NY: Suny Press.

———. (2010). The pedagogy of Momus technologies: Facebook, privacy, and online intimacy. *Qualitative Health Research, 20*(8), 1023–1032.

———. (2014). *Phenomenology of Practice*. Walnut Creek, CA: Left Coast Press.

van Manen, M., & Adams, C. (2009). The phenomenology of space in writing online. *Educational Philosophy and Theory, 41*(1), 10–21.

Verbeek, P.-P. (2005). *What Things Do: Philosophical Reflections on Technology, Agency, and Design*. University Park, PA: The Pennsylvania University Press.

———. (2008). Cyborg intentionality: Rethinking the phenomenology of human–technology relations. *Phenomenology and the Cognitive Sciences, 7*(3), 387–395. doi:10.1007/s11097-008-9099-x

———. (2016). Toward a theory of technological mediation: A program for postphenomenological research. In J. K. Berg, O. Friis, & R. C. Crease (Eds.), *Technoscience and Postphenomenology: The Manhattan Papers* (pp. 189–204). London, England: Lexington Books.

Vlieghe, J. (2015). Traditional and digital literacy: The literacy hypothesis, technologies of reading and writing, and the "grammatized" body. *Ethics and Education, 10*(2), 209–226.

———. (2016). Phenomenology of digital media. In M. A. Peters (Ed.), *Encyclopedia of Educational Philosophy and Theory* (pp. 1–6). Singapore: Springer. doi:10.1007/978-981-287-532-7_90-1

Chapter 2

Inviting and Interacting

Postphenomenology and the Microsociology of Education

Tobias Röhl

INTRODUCTION

Postphenomenology is a philosophical endeavor that promises to shed light on the ways in which humans and technology are entangled. In my ethnographic research on educational artifacts (Röhl, 2012b, 2013), I used postphenomological concepts in order to account for the materiality of education. Like other socio-material approaches, postphenomenology highlights that technological artifacts are not neutral tools but mediators of perception (Ihde, 1990) and action (Verbeek, 2005). They transform how and what we perceive and act in the world. In an "interrelational ontology" (Ihde, 2009, p. 44), artifacts and humans co-constitute each other. In different human-technology relations, artifacts amplify and reduce certain aspects of the world, but also invite some actions and inhibit others. They are sensually related to our body with its perception and abilities.

As I will argue in this paper, postphenomenology and its concepts enhance and sensitize microsociological and ethnographic accounts of education by opening up research for the invitational qualities of things and their bodily interrelations with human users without trivializing them. Drawing on ethnomethodological accounts of technology (Hindmarsh & Heath, 2000; Lynch & Macbeth, 1998; Greiffenhagen, 2014), this bodily interrelation is further combined with a practical interrelation in which the act of using things is seen as constitutive part of understanding technology. Consequently, I see technological mediation as a result of the interplay between invitational qualities of artifacts and their practical use, and argue for a further pragmatization of postphenomenology. This pragmatization builds on ethnographic observations of situated use of technology and extends postphenomenology's idea of "multistability" (Ihde, 1990, p. 144). For postphenomenology, technologies are "multistable"—that is, their purpose is not predetermined, but can vary in different contexts. While multistability is an important concept for postphenomenology, it is often something to be demonstrated (e.g., Ihde, 2009, p. 17ff.), rather than being used as an analytic starting point (Rosenberger, 2014, p. 381). As I will show, this analytic shift can be achieved by including the microsociological notion of situated use. What technologies are, for all practical purposes in a given situation, is also shaped by the varying way they are used and talked about; and this, in turn, displays to others how these artifacts are to be understood here and now.

In order to build my argument, I will first describe two types of engaging with educational artifacts that I encountered in my previous research. I will conclude my discussion by outlining methodological consequences of an approach combining postphenomenological concepts with microsociological research. While I make use of examples from the field of education, this approach can be applied to other fields as well.

WORKING THROUGH AND WITH THINGS: MEDIA AND EPISTEMIC OBJECTS

In the following, I will show how the invitational qualities of artifacts in human-technology relations mediate the ways in which teachers and students participate in education; and how this, in turn, enacts different types of artifacts. This enactment, however, does not solely rest on the invitational qualities of artifacts, but on the practical achievements of teachers, students and manufacturers: they work through and with things to enable their invitational qualities. The data presented here originates from ethnographic research on the role of artifacts in mathematics and science classes in German secondary schools (Röhl, 2013).

Transparent Media—Working Through Things

In education, a number of different media figure prominently in teaching and learning: blackboards, whiteboards (analogue and digital), notebooks, textbooks, projectors, televisions sets, video recorders, computers and tablets, and so forth. Some of them are established and commonplace tools of teaching (e.g., the blackboard); others can less frequently be encountered in the classroom and made their educational appearance only recently (e.g., PowerPoint presentations; see Adams, 2008). They are used to convey and distribute academic knowledge by different modalities (visual, auditory, semiotic). Some, like the blackboard, are reserved for the teacher; others, like the textbook, are distributed evenly among all participants. While some of them can be filled ad hoc with content in the classroom (e.g., notebooks, the blackboard), others have to be prepared at home (e.g., PowerPoint presentations), or produced beforehand at a publishing house (e.g., textbooks) or manufacturing company (e.g., demonstration experiments). Some can be used to address the whole class at once (e.g., demonstration experiments, whiteboards), others can only be used by a few persons at a time (e.g., textbooks, hands-on experiments).

Despite these differences, all these various artifacts can be described as media: they distinguish between a thematic foreground and a background—that is, they let us see something differently, or make something absent present, while at the same time repressing some features (Krämer, 2015; Wiesing, 2010). As such, they are part of "relations of mediation" (Ihde, 1990, p. 72ff.). They mediate the ways in which we perceive the world by becoming quasi-transparent. As "screens" (Introna & Ilharco, 2006), many educational media hold our attention and frame what is shown as relevant. While we focus our attention toward something that is shown, we are not concerned with the act of showing and how it is brought about (Debray, 2004). Like Heidegger's equipment (*Zeug;* 1962, pp. 66–88), media are "ready-to-hand" and thus not themselves objects of our immediate attention. Instead, we turn our attention toward the things made present by media.

This "hermeneutic transparency" (Ihde, 1990, p. 82) of educational media is the result of choices made by manufacturers and school administrators before class starts. This can be illustrated by taking a close look at one particular blackboard in a German school (see Figure 2.1). Since its surface is painted in dark green, the blackboard in German schools is actually a

Figure 2.1 A blackboard in a German school run by the Catholic Church (Tobias Röhl).

"greenboard." The dark and inconspicuous color of the surface contrasts with traces of white chalk applied to it and does not attract attention when the blackboard is not in use. The green surface is encased in a metal frame. Like other frames, it clearly distinguishes between an inside and an outside, thus denoting a relevant subject matter. Paraphernalia, like a cleaning cloth, can be placed in a tray beneath the frame. And the blackboard resides literally in the background of the classroom, fastened to the wall, towards which all students are aligned at their desks and chairs. Without much effort, they can look at it and see what is written on its surface. At the same time, the teacher is positioned next to it at his or her desk. While the blackboard behind him or her, and thus visually less accessible, the teacher can easily write on it. Two groups are spatially positioned in a way that facilitates specific ways of using the blackboard: viewing from a distance (students), and writing in proximity (teachers). By reducing any obstacles that may stand in the way of a "ready-to-hand" use, the blackboard can quite effortlessly serve as a medium in which some things (its surface, our bodily positions towards it, some of its components) can be relegated to the thematic background while other things come to the thematic foreground (the writing on its surface).

But how do teachers and students achieve this in the classroom? Using the blackboard as an example, I will show how educational media are made in the classroom, and become part of quasi-transparent relations of mediation. The blackboard has been a constitutive part of classroom teaching in many Western countries for roughly 200 years (Kidwell, Ackerberg-Hastings, & Roberts, 2008; Hamilton, 1990, p. 75ff.). With the advent of classroom teaching in the 19th century, it was established as a focal point of classroom discourse. Classroom discourse typically revolves around "elicitation sequences" (Mehan, 1998, p. 121) in which a teacher's initiation (usually a question) is met by a student's response, followed by an evaluation of that response through the teacher. This seemingly dialogic sequence is always directed at the whole class. With the blackboard, this sequence is further affirmed by creating a collective focus for the teacher's initiation; and also by creating a place where correct responses by students can be recorded, and thus turned into authorized knowledge through the teacher's hand, equipped with chalk.

Teachers and students direct their attention toward the blackboard frequently and it figures prominently in classroom discourse (see also Kalthoff & Röhl, 2011). Students are usually copying what is written on it into their notebooks, without any prompts by their teachers. Instead, teachers often have to remind their students that they have to wait before they are supposed to write something down if they want to elaborate an argument or develop a formula together with the class: "You don't need to write that down!"; "Pencils down and close your notebooks! There is something we need to think about!"; "Please wait till you write it down!" (classroom observations;

Röhl, 2013, p. 95ff.). This illustrates that copying is the prevalent activity in the classroom established around and with the blackboard. Especially in lower grades, students often aim to make an exact copy of the writings on the blackboard as evidenced by questions such as "Is that a heading?" "Are we supposed to colour that as well?" (classroom observations; Röhl, 2013, p. 95ff.). While it is an important part of the activities of teachers and students, the blackboard itself remains silent and rather passive. Teachers and students usually do not talk about the blackboard itself, but about their activities directed at it and the results of these activities—writing *on* the blackboard is in its double sense (activity, result) in the thematic foreground of the participants of school lessons: "I'll write down the important stuff you need to learn on the blackboard"; "I'm writing the example down"; "Open your notebooks and copy that!" (classroom observations; Röhl, 2013, p. 95ff.). That teachers and students direct their attention toward what is written on the blackboard, but not the blackboard itself, is taken for granted during class. Indeed, as a researcher I was often busy writing down what was on the board and forgot to take notice of how this was supported by the blackboard and how this affected students. I was similarly affected by the blackboard's qualities as a medium and reacted accordingly. Outside of class (e.g., during breaks), the blackboard can, however, easily become something else. The typical blackboard used in German schools with its foldable wings can, for example, serve as a hideout from other students, or as a cover from thrown objects like paper balls. It is then no longer an academic medium but an object used as a tool for something else.

By writing on the blackboard, teachers do not simply list a number of propositions or "hard" facts, but create a meaningful order.[1] Texts are neither written nor read in a linear fashion (Engert & Krey, 2013; Krämer, 2003). Instead, different textual elements (headings, paragraphs, text boxes, sketches, diagrams, formulas, etc.) are distinguished from each other typographically (by size, underlining, coloring, framing), and spatially related to each other (below/above, left/right, middle section/side sections). Teachers work on a semantically charged field in which different textual elements lend meaning to each other. Certain elements (important formulas, rote definitions) are singled out and thus highlighted in their relevance, by the teachers not only verbally stating their importance, but also by marking them visually with lines and colors. Other elements are relegated to less conspicuous places, like the side sections—for example, auxiliary calculations, homework assignments, administrative dates, and so on. The middle section usually signifies the main narrative of a lesson, and so on. What is written on the blackboard thus becomes meaningful, not by simply being there, but as a result of the participants' actions in the classroom. By writing something down in a specific place (e.g., as a headline), but also by referring to it verbally and gesturally, teachers often signify certain sections as important and assign specific disciplinary meaning to them: "What is important *here* [pointing to a formula]?"; "*That* [pointing to formula] is a differential equation!"; "*That* [pointing to x-axis] is what we call x, and *that* [pointing to y-axis] is what we call y." Students similarly try to make sense of the blackboard and assume that there is a meaningful order: "Why is there a 2 in parentheses?"; "Is *that* [pointing to the upper half of the blackboard] a heading?" (classroom observations; Röhl, 2013, p. 95ff.).

The blackboard and its writings also frame what happens in the classroom. Fickle classroom discourse is transformed into reproducible writing. Headings on the blackboard often serve as a reminder of what is talked about. Inappropriate remarks of students are postponed or rejected by referring to the heading on the blackboard. Therefore, a "public curricular field" (Macbeth, 2000, p. 36) is created that establishes a focal point for all participants. As soon as the lessons start, everything on the blackboard is thus a meaningful place. Even the empty blackboard is just a canvas for classroom discourse that has yet to be filled. And everything left empty can either later be written on or is left blank intentionally, and thus itself meaningful. By writing on the blackboard, teachers not only leave traces of chalk but also distinguishable empty spaces that signify something:

> The experiment is sketched on the blackboard [see Figure 2.2]. One student asks about the "gap" between the two electric fields, "Querfeld" and "Längsfeld," in the sketch: "The gap, is it intentional?" The teacher affirms this. He detaches the component from the experimental set-up and shows it to the class, so that they

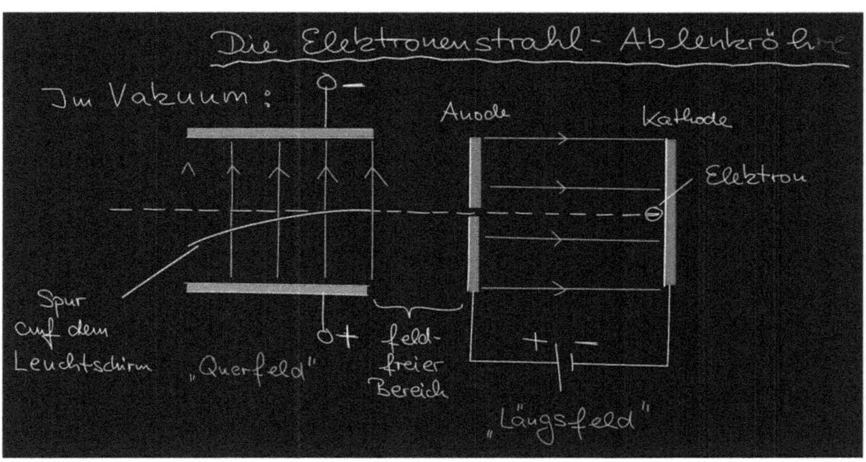

Figure 2.2 A meaningful gap on the blackboard (Tobias Röhl).

can see the "gap" at the device. Finally, he labels the "gap" on the blackboard: "feldfreier Bereich" [field-free area]. (classroom observations, science class, grade 12; Röhl, 2013, p. 100)

This example not only shows the relevancy and meaning of "gaps" but also another transparency of the blackboard. Things that are turned into writing on the blackboard, like the device in the example, are identified with the signs representing it. Often, teachers and students point to and talk about graphical inscriptions, formulas, and names on the blackboard, while at the same time referring to the actual objects which these represent. Even when they are absent, past experiments and objects shown a long time ago can be made present for the participants by evoking their written form.

The blackboard is thus enacted as an educational medium by designing its surface non-conspicuously, placing it at a focal point of the classroom, and by the participants focusing on the writing on its surface. A large and ever-present artifact is thus turned into a quasi-transparent medium by the efforts of several actors involved: educational manufacturers, school administrators, teachers, and students. They all work through this artifact to establish it as medium. Writing on the blackboard (both as activity and result) thus turns a void surface into a meaningful plane.

Epistemic Objects—Working with Things

Another important educational technology can be found in objects which are used to present knowledge themselves: in mathematics classes, models display features of geometrical bodies; and in science classes, experimental set-ups are used to show the laws of nature at work. These "epistemic things" (Rheinberger, 1997) differ from educational media, like the blackboard, insofar as they are not simply re-presenting curricular knowledge but presenting and embodying it (Mitchell, 1990; see also Kalthoff & Falkenberg, 2008). Consequently, a geometrical model is not just a sign but is indeed the geometrical body it re-presents. And experimental set-ups similarly not only stand for the laws of nature, but also demonstrate them at work. In terms of human-technology relations, we can classify the connections between these epistemic objects and human subjects as "alterity relations" (Ihde, 1990, p. 97ff.). We encounter something, as autonomous beings, that is independent from us and captivates our attention by its otherness.

Again, we can identify features that lend these objects to such relations. There are a number of manufacturers that produce epistemic objects for educational purposes. These objects all share a common feature—they are reduced to their essential and most relevant characteristics with regard to their educational purpose. If we take a close look at one of these objects, this becomes clear (see Figure 2.3). This model of a geometrical prism contains three pyramids. The prism is used to teach students the formula needed to calculate the volume of pyramids. By comparing the height and the base area of the pyramids to that of the prism, one can demonstrate that the formula is exactly one-third of the formula for prisms. While many different things can be described as geometrical prisms, one of the teachers I observed used this model and not some box. What makes the model better suited for educational purposes than household items and other things? The geometrical model is a purified object that is entirely devoted to its use as an educational artifact, demonstrating geometrical features of prisms and pyramids. There are no signs on its surface, it does not resemble anything else, but stands for itself. Its transparency lets one see inside, and thus the three pyramids it contains. The pyramids are clearly distinguished by their color—indeed, the sole purpose of the different colors is to quickly identify the shapes of three individual pyramids. Consequently, one can say that the purification and reduction of the object serves two purposes: First, from the perspective of the teacher, it can easily be used as a tool to teach and is—in this regard—transparent; students do not get distracted by features that are irrelevant to geometry (other purposes of an object, signs that invite

Figure 2.3 Geometrical model of a prism containing three pyramids (Tobias Röhl).

interpretation etc.). Second, by focusing on educationally relevant features they become at the same time amplified; these features are thus conspicuously presented as something which confronts us. It is this conspicuous embodiment of relevant features that turns epistemic objects, like the prism and its pyramids, into something opposed to us as subjects.[2] When one of the teachers I observed used the model of an airplane to show the students that one can discern its volume by assuming it is cylindrical in shape, he had to put much more effort and time into the transformation of an artifact into an epistemic object of geometry. In contrast to the prism, the airplane model is endowed with a number of qualities that distract from its status as cylindrical object: it is painted like its real-world counterpart, there are logos and writings on it, it has windows, wings, engines, and so forth.

How do students and teachers work with epistemic objects? In order to answer this question, I summarize some of my classroom observations of mathematics and science classes in German secondary schools in the following paragraphs (Röhl, 2013, p. 65ff.): Before epistemic objects enter the classroom their use is often rehearsed. In the case of experimental set-ups, teachers usually have to assemble different components and parts. They do this in preparation rooms where a number of different experiments and other educational tools are stored. After their assembly, teachers test if the experiment runs correctly and if it is clearly visible in front of a classroom full of students. When I observed them in the preparation rooms, the following questions where relevant for the teachers working on the experiments: Does the experiment work at all? Do the measurements adhere to the theoretical assumptions? Is the phenomenon visible from a distance? In their preparations they assure that a scientific phenomenon is visibly embodied by the experimental arrangement.

When the epistemic objects enter the classroom they are often placed at some place next to/at the side of the teacher's desk and not yet mentioned. Before the demonstration can start, a new topic has to be introduced and talked about. During this verbal introduction, I could observe that teachers turn their attention fully toward the class and ignore the experimental setup. Sometimes teachers have to assemble or adjust something, before they can begin with the experiment. Usually teachers do not comment these last-minute preparations and carry them out while the class is doing something else (e.g., writing something lengthy down). In one instance where a teacher commented on what she was doing, she kept her voice down and made clear that her activity was not part of the actual experiment but indeed preparatory: "This isn't part of the experiment yet!" (classroom observations; Röhl, 2013, p. 71). With their comments (or their omission) teachers signify preparatory activities as being profane and irrelevant acts beyond curricular knowledge.

In my field notes, teachers then often proceeded to place the experimental set-up prominently in the middle of the desk, making it the center of attention. Many even introduce the object explicitly by giving names to it or some of its parts:

> Now a little experiment! *This* [points to experimental set-up] is a basin filled with water, a 1-cent coin, and a copper pipe. (classroom observation, science class, grade 7; Röhl, 2013, p. 73)

> Alright, horizontal launch of a projectile, this would be something like that [goes to experimental set-up and tinkers with it]. We have here some kind of spring which you can tension [tensions the spring] and then I can release it and then it shoots somewhere [releases spring]. Fantastic, isn't it? (classroom observation, science class, grade 11; Röhl, 2013, p. 73)

By giving a name to them, both teachers in the above examples select some things. This serves two purposes. Firstly, elements are singled out to which the class can refer verbally. Secondly, these elements are assigned relevancy for disciplinary knowledge. Many parts of the experimental set-ups are not mentioned at all: clamps and screws may hold the set-up together and guarantee its execution, they are, however, not granted a name in classroom discourse. Nevertheless, students sometimes fail to see the named objects and their components in a purely disciplinary way—they, for example, identify (often jokingly) certain components with household items and let the class know what it looks like to them. Confronted with a blower that had a hose attached to it a student in one class exclaimed ironically: "Cool! A vacuum cleaner."

Finally, the experiment can be executed. Teachers prepare their students for the moment of execution and signify an adequate moment of observation: "Well, alright then!"; "Alright then. Ready, steady, go!" Often teachers also suggest how to observe the experiment: "Everyone look ahead!"; "Look carefully!"; "Pencils down! Look here!" (classroom observations; Röhl, 2013, pp. 79–80). The preferred sense is seeing—a visual event is to be observed. Elements that are not relevant are often explicitly excluded by stating their irrelevance: "This box just serves as a resonating body!"; "Unfortunately the acrylic glass is reflecting light as well, but this is not what it is about!" (classroom observations; Röhl, 2013, pp. 74, 81). Again, teachers shape how the experiment is perceived by distinguishing relevant and irrelevant elements and phenomena from each other. A "disciplinary vision" (Kalthoff & Röhl, 2011) is practically enabled in which things are to be seen from the perspective of an academic discipline. Seeing something as something is thus a practical achievement of the participants (see Goodwin, 1994; Lynch & Macbeth, 1998; Schindler, 2009). After the experiment has been conducted, teachers and students transform their observations into linguistic signs—both verbally and in writing on the blackboard and in the notebooks of students. This is the culmination of the disciplinary vision. The written formula on the blackboard only preserves a highly abstracted and generalized version of the demonstration—features that are seen as secondary to the experiment—like clamps and screws, colors, and so on, but also accidents and breakdowns—are, however, left out. A present thing is turned into a representation of selected features. By seeing disciplinary, some things are lost (clamps, screws, accidents, etc.); others, however, are gained (being able to generalize and extend idiosyncratic observations to other cases, being able to calculate and predict, etc.) (see also Latour, 1999).

The alterity of epistemic objects is highly manufactured and tamed. Science classes do not aim to let the things speak for themselves or to return to the things of the lifeworld. Instead, they can be seen as an attempt to discipline things in two ways—by making them work orderly and by making them part of a knowledge order of an academic discipline. Teachers thus work on a very specific alterity: phenomenon and thing become one in order to act as an independent entity that stands in contrast to other things and the human participants in the classroom; yet, they are restricted in their independence granted to them. Sooner or later, the otherness of epistemic objects is turned into a predictable and calculable formula.

METHODOLOGICAL CONSEQUENCES

What are methodological consequences of such an approach to (educational) technologies? In an earlier article, I outlined three general analytic strategies or stances that let us tackle the materiality of education (Röhl, 2012a; see also Adams & Thompson, 2011). Here I want to show more concrete methods that allow us to deal with these issues. In the previous section, different methods were implicitly employed to show the interplay of invitational qualities of educational artifacts and the activities connected to them.

Contrasting and comparing is one of the main operations of qualitative social research (e.g., Corbin & Strauss, 2008). By comparing data with other data, one can gain insights about the specific quality of given data material. It also enables an analytic perspective in which both the idiosyncrasies and the taken-for-granted nature of one's data are irritated. Ihde similarly argues for a pragmatic take on phenomenology's eidetic variation (2009, p. 12ff.). Instead of searching epistemologically for invariable essences, postphenomenology is interested in pragmatic and empirical variations of "multistable" (Ihde, 2009, p. 16) artifacts. Technological artifacts are not endowed with an essence that determines their use once and for all. They are developed in different directions, used in different ways, and so on. By comparing these different developments and uses and by systematically varying features of artifacts one can, however, gain insights into the bodily relations that a technology opens up—either on the level of perception ("amplification"/"reduction"; Ihde, 1990) or on the level of action ("invitation"/"inhibition"; Verbeek, 2005). Often, this is achieved by discussing examples from the history of technology (e.g., Ihde, 2009) or new technological developments (e.g., Verbeek, 2011). In both cases, however, this takes on the form of case studies that tend to emphasize general features of a technology and its use. While this approach avoids the pitfalls of romanticizing technology and analyzing technology as an abstract attitude characteristic of modernity—prevalent, for example, in Heidegger's accounts (Ihde, 1993)—it neglects the microsociological analysis of socially situated use, and how this use frames what things are to human actors and thus what can and cannot be done with them.

Consequently, I would like to take up the idea of variation and develop it into a tool for microsociological ethnographic research. Variation, then, is not something undertaken solely on the level of different (often historical) examples of technologies, but as something that can be found in ethnographic data of technology in use. From a sociological viewpoint, this has the advantage of being grounded in data that demonstrates how technology is actually used. Microsociological accounts of technology show that we are not determined by technology and that their use is not a secondary phenomenon. Instead, the context of use is seen as highly relevant, since it is here where things receive their practical meaning: "the sense and significance of objects emerge within the developing course of action and interaction; their objective and determinate sense is intersubjectively and momentarily accomplished 'here and now'" (Hindmarsh & Heath, 2000, p. 558). Things are worked on and change their status accordingly for the actors involved—for example, by pointing toward and talking about the screens in front of them, co-workers in a telecommunication control center create a shared definition of what they see. By turning variation into an analytic tool to investigate ethnographic data, one can account both for this situated use and the invitational qualities of artifacts. This *empirical variation* can assume different shapes. In the preceding sections, the following methods were used: (1) maximally contrasting artifacts, (2) minimally contrasting artifacts, (3) contrasting contexts of use, and (4) auto-ethnographic observations.[3]

Maximally Contrasting Artifacts

By comparing and contrasting two types of different artifacts it is possible to gain insights into the specific features of a given technology. In the case of educational artifacts, one can easily classify quasi-transparent media and epistemic objects as two different categories. They belong to different categories not just by being designed differently but also by being used in fundamentally different ways. Media, like the blackboard, are used to represent something which, in turn, is talked about, copied, and so on. The medium itself is absent from the thematic foreground of participants' actions and efforts. By treating the blackboard as something which does its job in the background, and by putting effort into its transparency, it is indeed performed as a medium with a meaningful plane on which to operate. Epistemic objects, on the other hand, are worth talking about from the perspective of teachers and students. For them, they embody knowledge and are thus treated as being endowed with a fascinating "presence" (Gumbrecht, 2004) of an object in its otherness. By making them the focal point of their activities, participants turn them into something worthy of taking center stage. And at the same time, these artifacts are enacted as objects in the narrow sense—they are seen as something independent from us and our activities. Ultimately, we can identify two objects that are made differently at two points: in development/design and in actual use. At the same time, these analyses show that there are also overlaps between these two types of artifacts.

Minimally Contrasting Artifacts

Another way of contrasting artifacts is to take two artifacts that are used for similar purposes. This is an operation of minimal contrast in which one can dig deeper into the invitational qualities of technology. How can one explain the different ways that an artifact is used? Why is an artifact designed in a specific way? How do different qualities contribute to its actual use? In the case of educational technologies, I could show that artifacts that are specifically designed for educational purposes are more easily integrated into classroom discourse. This was achieved by contrasting the geometrical prism and the airplane model. In doing so, features of an educational design could be identified: the educational purification and disambiguation of artifacts.

Contrasting Contexts of Use

Technological artifacts are not determining practice, but inviting different ways of using them. This is one of the insights of postphenomological thought. This is also in line with practice theory as envisioned by Schatzki (2002). How artifacts are used is not a question of enforcement but one of adequacy and appropriateness (Schatzki, 2002, p. 210ff.). Educational artifacts, in particular, invite us to use and see them disciplinarily, there are no "scripts" (Latour, 1994) that enforce a certain way of using them. To put it differently: the artifacts described here are not Latourian speed bumps, but open to re-interpretation and re-purpose. When teachers and students work on things to turn them step-by-step into epistemic objects, they dynamically change them. Something which attracts attention by being there and behaving unexpectedly is finally tamed by transforming it into signs on the blackboard. And when students use the blackboard as a prop in their peer cultural activities and direct their attention toward the thing itself, we can see more clearly how, in contrast, it is used as a medium. Or when students see epistemic objects not as educational artifacts but as household items, then we can similarly glimpse the efforts that go into enacting them as embodiment of a disciplined otherness. In that sense, contrasting contexts of use is another way of exploring the postphenomenological notion of multistability. Yet, it extends this notion by looking for this multistability in ethnographically observed situations where a particular artifact was used.

Auto-ethnographic Observations

The last method I want to discuss is not about contrasting data. Instead it relies on a reflective attitude in which the researcher becomes the object of his or her own observations. This can be achieved by either focusing on one's own reactions and writing them down, or by using previously recorded data and asking about the taken-for-granted assumptions therein. This was done by analyzing how different ways of observing are required when we encounter either the blackboard or the experiments in the classroom. In the case of educational media we need to turn our attention toward features that the participants themselves are ignoring. By assuming such an analytic stance from time to time in the field, we can make features visible to research that are usually taken for granted. This holds especially true for socio-material research where we have to challenge our everyday modernist assumption, in which objects are mere tools used by human subjects, both in their separate worlds (Latour, 1993). Additionally, both phenomenology (e.g., Heidegger, 1962) and pragmatism (e.g., Mead, 1972) teach us that objects reside in the thematic background when we use them, thus becoming inconspicuous. Consequently, we need an analytic stance that deviates from those of the people we are observing if we are interested in the pragmatic knowledge surrounding technology and its use.

In employing these methods I could show how a socio-material and microsociological account of education can benefit from postphenomenological concepts like mediation, human-technology relations and variation. This opens up microsociological research for the sensual qualities of artifacts and the bodily relations between them and their users. Artifacts are not solely shaped by their use, but have qualities beyond that (Harman, 2009). Postphenomenology, on the other hand, can gain a better insight into the pragmatic and situated side of technological mediation. Human-technology relations are worked upon and brought about by actors engaging in them. This also highlights that human-technology relations are not clear cut categories, but denote a dynamic range in which we have to constantly situate ourselves in face of artifacts surrounding us and our practice. The alterity of epistemic objects in education, for example, is a practically controlled and situatedly specified otherness that deviates from that of the toys and robots described by Ihde (1990). And while microsociology extends its scope beyond human actions, postphenomenological analyses are validated and further elaborated upon through ethnographic and microsociological research. Socio-material practice can thus ultimately be described as interplay between invitational qualities of artifacts and the way in which we interact with them.

NOTES

1. For the field of university mathematics Greiffenhagen (2014) similarly observed that by writing on the blackboard an "*organized* body of knowledge" (513) is brought about.

2. In that regard educational epistemic objects are also media: some features are highlighted (e.g., geometrical features), others are inhibited (e.g., mundane associations). Both (media and objects) bring some things into the thematic foreground, and others into the thematic background. In the case of educational epistemic objects, however, what is brought into the foreground is present in the object itself—and not another re-presentation (like the writing on the blackboard). One could also argue that educational epistemic objects rest on qualities that are more akin to "technical things" (Rheinberger, 1997): they have qualities that make them reliable tools (*Zeug*; Heidegger, 1962, pp. 66–88) to display other qualities that raise questions and are experienced as something independent from us and opposed to us.

3. In a similar vein, Rosenberger (2014) describes how "variational cross-examination" can be used to extend the postphenomenonological notion of multistability. By brainstorming different possibilities for the design of the analyzed artifact, its use, and its relation to other human and nonhuman actors one can outline what makes specific human-technology relations stable (or unstable). In contrast to this quasi-eidetic approach, my methods rely on ethnographic observations and comparisons. Nevertheless, both approaches acknowledge that situated use contributes to the stabilization of objects.

REFERENCES

Adams, C. A. (2008). PowerPoint's pedagogy. *Phenomenology & Practice, 2*(1), 63–79.

Adams, C. A., & Thompson, T. L. (2011). Interviewing objects: Including educational technologies as qualitative research participants. *International Journal of Qualitative Studies in Education, 24*(6), 733–750.

Corbin, J. M., & Strauss, A. (2008). *Basics of Qualitative Research: Techniques and Procedures for Developing Grounded Theory*. London: Sage.

Debray, R. (2004). *Transmitting Culture*. New York, NY: Columbia University Press.

Engert, K., & Krey, B. (2013). Das lesende Schreiben und das schreibende Lesen: Zur epistemischen Arbeit an und mit wissenschaftlichen Texten. *Zeitschrift für Soziologie, 42*(5), 366–384.

Goodwin, C. (1994). Professional vision. *American Anthropologist, 96*(3), 606–633.

Greiffenhagen, C. (2014). The materiality of mathematics: Presenting mathematics at the blackboard. *The British Journal of Sociology, 65*, 502–528.

Gumbrecht, H. U. (2004). *Production of Presence: What Meaning Cannot Convey*. Stanford, CA: Stanford University Press.

Hamilton, D. (1990). *Learning About Education: An Unfinished Curriculum*. Buckingham: Open University Press.

Harman, G. (2009). *Prince of Networks: Bruno Latour and Metaphysics*. Melbourne: Re.Press.

Heidegger, M. (1962). *Being and Time*. New York, NY: Harper & Row.

Hindmarsh, J., & Heath, C. (2000). Sharing the tools of the trade. The interactional constitution of workplace objects. *Journal of Contemporary Ethnography, 29*(5), 523–562.

Ihde, D. (1990). *Technology and the Lifeworld: From Garden to Earth*. Bloomington, IN: Indiana University Press.

———. (1993). Deromanticizing Heidegger. In D. Ihde (Ed.), *Postphenomenology: Essays in the Postmodern Context* (pp. 103–115). Evanston, IL: Northwestern University Press.

———. (2009). *Postphenomenology and Technoscience. The Peking University Lectures*. New York, NY: State University of New York Press.

Introna, L. D., & Ilharco, F. M. (2006). On the meaning of screens: Towards a phenomenological account of screenness. *Human Studies 29*(1), 57–76.

Kalthoff, H., & Falkenberg, M. (2008). Kommunikation unter Anwesenden: Lehrer—Schüler—Medien. In H. Willems (Ed.), *Lehr(er)buch Soziologie. Für die pädagogischen und soziologischen Studiengänge* (pp. 909–930). Wiesbaden: VS.

Kalthoff, H., & Röhl, T. (2011). Interobjectivity and interactivity. Material objects and discourse in class. *Human Studies, 34*(4), 451–469.

Kidwell, P. A., Ackerberg-Hastings, A., & Roberts, D. L. (2008). *Tools of American Mathematics Teaching, 1800-2000*. Baltimore, MD: Johns Hopkins University Press.

Krämer, S. (2003). Writing, notational iconicity, calculus: On writing as a cultural technique. *Modern Languages Notes—German Issue, 118*(3), 518–537.

Krämer, S. (2015). *Media, Messenger, Transmission. An Approach to Media Philosophy*. Amsterdam: Amsterdam University Press.

Latour, B. (1993). *We Have Never been Modern*. Cambridge, MA: Harvard University Press.

———. (1994). Where are the missing masses? The sociology of a few mundane artefacts. In W. E. Bijker & J. Law (Eds.), *Shaping Technology / Building Society* (pp. 225–258). Cambridge, MA: MIT Press.

———. (1999). Circulating reference: Sampling the soil in the Amazonas forest. In B. Latour (Ed.), *Pandora's Hope: Essays on the Reality of Science Studies* (pp. 24–79). Cambridge, MA: Harvard University Press.

Lynch, M., & Macbeth, D. H. (1998). Demonstrating physics lessons. In J. G. Greeno & S. V. Goldman (Eds.), *Thinking Practices in Mathematics and Science Learning* (pp. 269–297). Mahwah, NJ: Erlbaum.

Macbeth, D. H. (2000). Classrooms as installations: Direct instructions in the early grades. In S. Hester & D. Francis (Eds.), *Local Educational Order: Ethnomethodological Studies of Knowledge in Action* (pp. 21–71). Amsterdam: John Benjamins.

Mead, G. H. (1972). *Philosophy of the Act.* Chicago, IL: University of Chicago Press.

Mehan, H. (1998). The structure of classroom discourse. In T. A. van Dijk (Ed.), *Handbook of Discourse Analysis III: Discourse and Dialogue* (pp. 120–132). London: Academic Press.

Mitchell, W. J. T. (1990). Representation. In F. Lentricchia & T. Mc Laughlin (Eds.), *Critical Terms for Literary Study* (pp. 11–22). Chicago, IL: University of Chicago Press.

Rheinberger, H. J. (1997). *Toward a History of Epistemic Things: Synthesizing Proteins in the Test Tube.* Stanford, CA: Stanford University Press.

Röhl, T. (2012a). Disassembling the classroom: An ethnographic approach to the materiality of education. *Ethnography and Education, 7*(1), 109–126.

———. (2012b). From witnessing to recording: Material objects and the epistemic configuration of science classes. *Pedagogy, Culture & Society, 20*(1), 49–70.

———. (2013). *Dinge des Wissens. Schulunterricht als sozio-materielle Praxis.* Stuttgart: Lucius & Lucius.

Rosenberger, R. (2014). Multistability and the agency of mundane artifacts. From speed bumps to subway benches. *Human Studies, 37*(3), 369–392.

Schatzki, T. R. (2002). *The Site of the Social: A Philosophical Account of the Constitution of Social Life and Change.* University Park, PA: Pennsylvania State University Press.

Schindler, L. (2009). The production of "Vis-Ability": An ethnographic video analysis of a martial arts class. In U. T. Kissmann (Ed.), *Video Interaction Analysis: Methods and Methodology* (pp. 135–153). Frankfurt a.M.: Peter Lang.

Verbeek, P.-P. (2005). *What Things Do: Philosophical Reflections on Technology, Agency, and Design.* University Park, PA: Pennsylvania State Press.

———. (2011). *Moralizing Technology: Understanding and Designing the Morality of Things.* Chicago, IL: University of Chicago Press.

Wiesing, L. (2010). *Artificial Presence: Philosophical Studies in Image Theory.* Palo Alto, CA: Stanford University Press.

Chapter 3

Entering the Portal

Media Technologies and Experiential Transportation

Jesper Aagaard

In *How We Became Posthuman* (1999), N. Katherine Hayles writes that "when people begin using their bodies in significantly different ways, either because of technological innovations or other cultural shifts, changing experiences of embodiment bubble up into language, affecting the metaphoric networks at play within culture" (p. 206). Arguably, one of the most significant cultural shifts of the 21st century has been the increasing influx of digital technologies. The contemporary landscape is characterized by what has been called an invasion of glowing rectangles (McCullough, 2013). But what are the repercussions of this digital colonization of the human lifeworld? How, for instance, does it affect our everyday experiences of space and time? And how is it best embedded in what Hayles calls "metaphoric networks"? In my research on technological mediation of attention, I have long looked for ways to understand the phenomenological role of media technologies in our lives. In this endeavor, I stumbled upon Galit Wellner's (2011) promising post-phenomenological metaphor of a *wall-window*. By comparing this metaphor to my own empirical material as well as my experience as a technologically equipped citizen, however, I eventually came to disagree with its radical conceptualization of technologically mediated attention.

The purpose of this chapter is to sketch out an empirically informed alternative metaphor for media engagement. Based on empirical material from a qualitative study in a Danish business college, I argue that we can understand engagement with media technologies in terms of another architectural metaphor, namely that of a *portal*. This metaphor entails a specific understanding of technologically mediated attention: whereas the wall-window metaphor leads to optimist ideas about so-called non-reductive attentional splitting between physical and virtual space, the portal suggests that this kind of flawless multitasking is in fact impossible. The rest of the chapter is structured as follows: First, I flesh out the wall-window metaphor in terms of its meaning and its unorthodox notions of attention. I then point out what I perceive as flaws in the metaphor and use illustrations from my empirical research to argue that we should instead conceptualize our engagement with media technologies in terms of a portal, which opens up to considerations of absence and presence. I then go on to discuss methodological issues of interest to an empirically informed postphenomenology. Finally, I address the ethics of attention implied by the metaphor of the portal.

THE WALL-WINDOW: NON-REDUCTIVE ATTENTIONAL SPLIT

When using a technology, there are multiple ways we can relate to it. It is *multistable* (Ihde, 1977). This means a technological artifact does not have an essence, but in fact supports several coherent stabilities. Using this postphenomenological insight, Galit Wellner (2011) proposes that one stable way we relate to cellphones is as a *wall-window*: when using a cellphone, you erect an imaginary *wall* between yourself and your immediate circumstances (e.g., the train), while the phone provides a *window* to remote space (e.g., the space of your interlocutor). Wellner goes on to suggest that this wall-window metaphor can be extrapolated to all media technologies. In the case of the computer, for instance, the window then refers to the remote space of the Internet as seen through the screen. According to Wellner, the common denominator of all media technologies is their "capacity to *dislocate our attention* to other places (remote in place, time or imaginary), while our physical body remains in *full engagement* with the surroundings" (2011, p. 95; emphasis mine). As implied by the casual juxtaposition of "dislocated attention" and "full engagement," the wall-window rejects conventional notions of attention in which the more attention you give to virtual space, the less you give to your immediate surroundings, and vice versa.

Wellner presents the revolutionary possibility of a *non-reductive attentional split* in which attention is paid to both virtual space and one's immediate surroundings. By doing so, she effectively replaces the traditional zero-sum understanding of attention with a kind of win-win situation: "It is not an 'either-or' relation," she claims, "but an 'and-and' that covers the physical and the remote spaces" (2011, p. 97). A closer look at the wall-window metaphor thus shows that it does not abide by the normal semantics of "window": windows exist to allow ventilation and the passage of light into closed off buildings. Without walls, there is neither need nor place for windows. According to this logic, any time a person engages with a "window" to remote space, this person must already have erected a "wall" to his or her surroundings. This, however, is not how the metaphor is built. Instead, the wall-window is conceived as a unitary phenomenon comparable to "transparent floor-to-ceiling walls" (Wellner, 2011, p. 80). In other words, the wall-window is envisioned through the language of the gaze in which a user may simultaneously peek through a transparent wall and a window. Wellner illustrates such non-reductive attentional splits through descriptions of teenagers who do their homework with the TV on and so-called *persistent phone users* who continue to perform here-and-now activities while engaging with their phones.

It is worth noting that this idea of a non-reductive attentional split closely coincides with popular notions of young people as *digital natives* (Prensky, 2001) or a kind of *net generation* (Tapscott, 1998). According to these discourses, frequent interactions with media technologies have rewired the brains of today's youth and given them an almost innate ability to attend to multiple streams of information simultaneously. Indeed, Wellner (2014a) has recently argued that "we can re-wire our brains, because the brain is plastic, that is highly flexible in its ability to adapt to various forms of technologies, situations and experiences, such as celling while driving" (p. 6). In other words, the youth of today are considered superb multitaskers. As claimed in the now-famous and oft-cited anthology *Educating the Net Generation* (2005), "Net Gen students are facile at multitasking and moving back and forth (sometimes rapidly) between real and virtual spaces" (Brown, 2005, p. 176). As these new generations enter our schools, it is argued, we must reconfigure the educational system to accommodate to their new ways of learning: students will be bored without the rapid changes that media technologies can provide, and media multitasking engages them because it allows them to do several things at once. It is with this rosy picture of multitasking that I take issue. A plethora of scientific research shows that young people are not the superb multitaskers that the digital natives discourse portrays them to be (Aagaard, 2014). If teenagers do their homework with the TV on, for instance, it significantly impairs their homework performance (Pool, Koolstra, & van der Voort, 2003). The accompanying idea of "persistent phone use" is what this chapter sets out to challenge.

THE PORTAL: EXPERIENTIAL TRANSPORTATION

Using illustrations from my empirical work, I will argue that salient aspects of the experience of using media technologies are best conceived through the metaphor of a *portal*. Before presenting my empirical material, however,

I want to emphasize that I appreciate Wellner's (2011) attempt to flesh out the phenomenology of media technologies. In fact, the notion of a portal can be said to spring from the margins of her article: Wellner's (2011) descriptions of media use contain notions of being "transported" to a remote space through technology (p. 81), and of a texting student "returning" to class (p. 92); notions that can hardly be explained through images of transparent floor-to-ceiling walls. In an essay on Husserl entitled "The Philosopher and His Shadow" (1964), Merleau-Ponty cites Heidegger's remark that "the greater the work accomplished . . . the richer the unthought-of element in the work" (p. 160), and uses this idea to get at aspects of Husserl's work that pervade his work, but are so close to him that Husserl himself was unable to see them. In what follows, I use this cooperative strategy in an attempt to get at hitherto unthought-of elements in Wellner's writing.

Situating the Current Study

Let me begin by briefly situating my research project: The empirical material presented here is part of a broader study on technological mediation of attention in an educational context. The study is conducted as a long-term, multi-method qualitative inquiry at a large business college in urban Denmark. The fieldwork began in August 2013 and spanned a year and a half. After six months of open, ethnographic participant observation in various classrooms, I started formally interviewing individual students about their use of technologies. These interviews were semi-structured, meaning they took departure in an interview guide, yet remained flexible enough to explore spontaneously occurring "red lights" such as unusual terms or intonations in the participants' answers (Kvale & Brinkmann, 2008, p. 135). All participants volunteered and were not paid for any of their involvement. During my observations, I had noticed how students often used educationally unrelated websites during class (Aagaard, 2016a). Students reacted particularly strongly against any kind of perceived passivity, which they seemed to find almost unbearably boring (Aagaard, 2015). In an attempt to gauge whether this type of *mediated impatience* was a strictly educational phenomenon, I wanted to address the use of digital devices outside of school in settings that young people choose freely. A number of interview questions thus regarded the students' spare time ("How do you use technology outside of school?" "Do you use social media when you are with your friends?" etc.). In total, 25 students were interviewed. Interviews lasted about 15 minutes each and the sound recordings were subsequently transcribed to text. The quotations used below are those which best illustrate the points of interest. They have been translated from Danish to English and all names are pseudonyms.

Entering Another Realm

During an interview with a student, Jim, he told me that he would often go on Facebook or play online games during lessons he found boring. However, Jim also told me that now, as an ambitious senior, he was trying to limit this kind of media use, but administering this kind of self-discipline had proven to be difficult. Jim had basically trained himself to log onto Facebook during lessons, and breaking this habitual distraction had proven harder than he had thought. When you have a computer right in front of you, he explained, the pull toward social media is immense. This made me interested in Jim and his friends' use of phones and social media during everyday interactions.

JESPER: Do you also do that when you're together with friends? If there's a break, do you go on Facebook?
JIM: A lot of people have their phones right next to them and usually have something going on. They're either in a conversation, snapping [using Snapchat], or on Facebook. Sometimes I think it's annoying if you spend time with people and they're anti-social with their phones. So yeah, I'd say it happens.
JESPER: But you find it a bit "annoying," you say?
JIM: Yeah, I do. The way the phone has taken control of us. You can be somewhere completely different even though you're sitting together.
JESPER: And what is "annoying" about your friends doing that?
JIM: Well, you're hanging out and trying to have a conversation, so it's pretty provocative that they're somewhere else. You've come together to be together. So, you should really be together.

Jim explains that a lot of people use their phone to access social media, but this use has the effect of rendering the user completely unresponsive to their friends. As Jim describes it, it is as if the user is located somewhere completely different although he or she may be physically seated right next to you. It feels like the person has mentally exited the room and left behind only the empty shell of a body. At first glance, it is almost as if digital devices are capable of instantiating a paradoxical split between the person's body and mind (hence the opening quote). Upon closer examination, however, it becomes evident that media engagement involves manual comportment like clicking, scrolling, and swiping (Friesen, 2011). But although there is never talk of pure consciousness emanating from a body in suspended animation, co-present friends experience a discrepancy between a person's physical body and their attention, which is experienced as "annoying." In contrast to this movement away from ones social circumstances, Jim emphasizes the primacy of embodied situatedness: when you have shown up somewhere to be together with your friends, he argues, it is common courtesy to devote attention to those friends ("You've come together to be together. So, you should really be together."). Even if you might temporarily lose interest in what is being talked about, you hang in there. Simply transporting yourself somewhere else is "annoying," "anti-social," and "provocative." When I asked another student, Karen, whether she found it acceptable to partake in social media when hanging out with your friends, she mirrored many of these sentiments.

JESPER: Currently, there's big social debate on whether it's okay to go on Facebook when you're together with your friends. What do you think?
KAREN: I think it's very rude. When you talk to people, that they just barely listen to what you have to say.
JESPER: "Rude?" In what sense?
KAREN: You can tell that they're focused on something else. I believe you should be together with the people you're with. Not just sit there and do all sorts of other things. To me, it's like the person went into another room and couldn't even hear what I'm saying, because they're not paying attention anyway.

Continuing the normative description from above, Karen also finds it "rude" to use social media when spending time with other people. She also describes the situation in spatial terms: It is as if the person has gone into another room. Notice that Karen expresses herself in cryptic and seemingly tautological terms: "You should be together with the people you're with." How can we do anything but that, a skeptic might ask. But while this statement may not make much *logical* sense, it makes perfect *phenomenological* sense as a vivid description of attention and lack thereof. Like Jim, Karen describes how a person focused on something else feels distant, like he's not really there. Also like Jim, however, she describes these awkward encounters in terms of friends' departures. They both denounce other people's tendency to do annoying things with their devices, but seem to exempt themselves from culpability when describing such everyday scenarios. Their answers thus bring to mind the so-called social desirability bias, which is a tendency to answer questions in a manner that is viewed favorably by others. One possible explanation for these statements is that Jim and Karen are simply reproducing negative discourses about digital devices perpetuated by educators, parents, and the media. Against this idea I would emphasize that many students were acutely aware of their own use of devices. Far from merely regurgitating familiar tales, they provided reflexive and in-depth accounts of their own "annoying" and "rude" uses of technology. One example occurred in an interview with the student Kent, who started out by explaining that he gets annoyed if people scroll through Facebook while he is together with them. I asked him what about it that annoys him.

KENT: You always hear that men can't multitask. And this is a pretty good example that men can't multitask when they have a phone in their hand. You're just inhabiting two worlds. And apparently a lot of people can't cope with that. Sometimes I have my phone in my hands—this is something I'm quite bad at myself—but then, if my girlfriend says something to me I have to ask, "What?" two or three times. Then she gets annoyed with me and feels like she isn't interesting to listen to. But I'm not doing it on purpose. It's basically because I'm not aware that, "Okay, I'm now using my phone and may seem completely absent to my surroundings." So yeah.

As we see, Kent self-consciously shifts perspective mid-sentence and moves from a discussion of the misdeeds of his friends to his own complicated relationship to the phone. Although starting out by complaining about other people's tendency to use phones at inappropriate times, Kent ends up admitting to doing the same thing himself. He explains to me that "as a man" he cannot multitask. Instead of simultaneously inhabiting two worlds when using his phone during conversations, what we have previously called persistent phone use, Kent actually seems "absent" from his surroundings—much to the dismay of his girlfriend, who feels misunderstood and underappreciated when this happens. In this brief segment, Kent both acknowledges that there is a problem ("this is something I'm quite bad at myself"), and tells me that he is not doing it out of malicious intent ("But I'm not doing it on purpose"). What we may surmise, then, is that Kent has developed a habit that is not completely under his voluntary control (Aagaard, 2015). In summary, when engaging with media technologies, people can be moved to mediated realms of elsewhere. Such departure may to some extent depend on active intentionality on behalf of the user, but it leaves behind an empty shell of a body to spatially proximal companions. This is considered both annoying and rude. Hence, when confronted with empirical realities, the notion of persistent phone use seems to carry significant phenomenological omissions. Of course, these empirical excerpts stem from only a few interviews, and although the other students expressed similar sentiments, this does not preclude the existence of persistent phone use. However, I leave the question of post-phenomenological validity for a later section. Trusting that these excerpts are sufficient to make my point, I subsume these descriptions under the heading of a *portal*.

The Portal: Absence and Presence

The students use spatial language to describe the comportments of multitasking phone users: They feel like these friends are located somewhere completely different, like they have gone into another room, and they seem absent from their surroundings. Based on such descriptions, I would argue that the use of media technology is best conceived through the metaphor of a portal. The word "portal" means doorway, entrance, or passage and descends from Medieval Latin *portāle*, city gate. Phrasing media engagement in terms of entering a portal implies movement from one place to another and for our metaphorical purposes this means that media users are not sitting-inside-and-looking-out at mediated content, but actively entering and exiting specific realms. This experiential transportation is not a radically new phenomenon that only comes into existence with the introduction of screens, but can also be found in the familiar experience of reading books. According to Friesen (2011), reading, writing, and surfing the web are all ways to access "another realm of experience and meaning" (p. 83). This is not to say that the dynamics in these cases are identical (see, e.g., Mangen & Kuiken, 2014). The point is that by entering media portals, whether books or phones, we leave behind our immediate world and its inhabitants in favor of mediated realms. Our everyday language reveals the importance of such experiences: We can be absorbed, immersed, captivated, engrossed, or engulfed by something. We even speak of people being lost in a book. Hence, the portal inclines us to discuss engagement with media technologies in terms of *absence* and *presence* in particular places (e.g., at the dinner table, in a plot, in the classroom, or on Facebook). And because being present in two places at once is simply not possible, we are inclined to reject the notion of non-reductive attentional splitting. Let me be clear: I am not disputing that in certain circumstances ones *physical body* can be located in two places at once (e.g., at the Rainbow Bridge at Niagara Falls bordering Canada and the United States). What I am claiming is that a *person* is not fully present in two places at once when merely standing in the doorway or vacillating between them. This notion of presence is tied to a phenomenological description of the human condition as Dasein, being-there. On the basis of being-there, Martin Heidegger (1995) addresses the privative notion of being-away:

> How often it happens, in a conversation among a group of people, that we are "not there," how often we find that we were absent, albeit without having fallen asleep. This not-being-there, this being-away, has nothing at all to do with consciousness or unconsciousness in the usual sense. On the contrary, this not-being-there can be highly conscious. In such being absent we are precisely concerned with ourselves, or with something else. Yet this not-being-there is nonetheless a being-away. (Heidegger, 1995, p. 63)

When conversing with my friends I can be both physically present and highly conscious, but if my consciousness is directed elsewhere (e.g., the contents of my smartphone), it will literally seem to my friends as if I am "not-being-there." That this absence is experienced from the perspective of my friends is a crucial point, which Heidegger leaves too implicit. Judgments of absence and presence always occur from the second-person perspective, which corresponds to ethical concerns (Friesen, 2012). As van Manen (1990) argues, when a person says of his friend that she is lost in a book, then *he* is actually the one experiencing a loss, "namely the attentive presence of his companion," (p. 61). The notion of a portal thus implies a perspectival shift in discussions of multitasking: You might think you are capable of persistent phone use, but what would your friends say? This perspectival shift explains the Cartesian specter that seems to haunts this chapter: While there is no discrepancy between a person's body and their mind, a noticeable lack of movement and "vitality dynamics" makes it *seem that way* to others (Aagaard, 2016b).

Because presence is primordial, we do not ordinarily give much thought to the presence of others during everyday conversations, but we do notice when they are unresponsive and preoccupied with something else. This peculiar experience constitutes a breakdown in everyday Mitsein, being-with: We get "annoyed" or even hurt if we talk to somebody and it seems like they are not paying attention. We feel distant from a person who is spatially close, but mentally remote. Heidegger's reflections on being-away obviously predate our modern media technologies, but it does seem like contemporary artifacts such as smartphones have made the experience proliferate. People have even coined new terms such as "nocializing" and "phubbing" to address these issues: *Nocializing* is the act of being out in a public setting like a bar and spending time on a digital device instead of socializing with the people around you. *Phubbing* is a portmanteau of "phone" and "snubbing" and refers to snubbing someone in favor of a mobile phone whenever a conversation loses its appeal. Studies of media technologies need a language that is sensitive to such human experiences. The need for such expressions reveals something additional about the use of media technologies: While the intensely immersive experience of media portals like books and smartphones might be similar, the norms currently governing the use of such technologies are strikingly different. Personally, I have never experienced being snubbed mid-conversation in favor of a book. What I have experienced, however, is being snubbed in favor of social media or text messages and, full disclosure, have occasionally snubbed others for the same reasons. None of these cases contained non-reductive attentional splits. It seems more appropriate to discuss such events in terms of absence and presence.

Metaphors: Semiotic Mediation

"The wall-window may be an unfortunate metaphor," responds the skeptic, "but why rectify a mere metaphor?" Metaphors are usually regarded as handy shortcuts to convey complex ideas. They are instruments of thought that explain novelty via familiarity (Radman, 1997): Time is money. Life is a journey. The mind is a computer. Each metaphor powerfully illuminates and enriches a specific idea. But metaphors are also encumbered with assumptions, and when people use metaphors, they inadvertently embed those assumptions into the conversation. The prominent conceptual metaphor "argumentation is war," for instance, structures the way we tend to argue and discuss. We *defend* a standpoint, *attack* our opponent's views, get *shot down*, *lose* the argument, and so on (Lakoff & Johnson, 1980). The metaphor emphasizes the combative aspect of arguing and once this warlike relation has been established, it becomes hard to imagine (much less conduct) arguments in cooperative terms. Hence, metaphors are not just neutral rhetorical devices that reflect a preexisting reality; they pragmatically affect how we experience and handle everyday situations. Unfortunate metaphors may lead to socially harmful or undesirable framings of certain phenomena (Schön, 1993). This tangible influence on our perceptions, thoughts, feelings, and actions can be considered a kind of *semiotic mediation* (Brinkmann, 2014b). Semiotic mediation is the discursive counterpart to the postphenomenologically familiar notion of technological mediation. Stating that a specific type of comportment qualifies as "non-reductive attentional splitting," for example, implies a different set of understandings and possibilities compared to calling it "absence." What's more, such metaphors shape laws and policies about multitasking and media use. The notion of digital natives is already driving the debate about social, political, and educational legislation (Bennett & Maton, 2010). Since the wall-window is so perfectly aligned with idealist discourse on young people's skilled use of technology, I find it urgent to identify and circumvent its inherent limitations to avoid potentially misguided decision-making. As Richardson and Wilken (2012) argue, "The ontological and cultural significance of the window and the frame, especially for understandings of media, cannot be overstated" (p. 186).

POSTPHENOMENOLOGICAL METHODOLOGY

In laying forth my empirical material, I am not advocating an incorporation of some sort of naïve positivism into postphenomenology nor proposing that our discussions turn into closed questions answered only by empirical material. I am just trying to open up the field by encouraging empirical explorations of other people's technologically mediated experiences. Since this is still a relatively new endeavor in the field of postphenomenology, I will discuss the two important methodological notions of *researcher reflexivity* and *analytical validity* from this theoretical perspective.

Postphenomenological Reflexivity

When it comes to researcher reflexivity, empirical branches of phenomenology have traditionally been built around the key Husserlian epistemological strategy of suspending ones natural attitude in an attempt to remain open to the "things themselves" in their appearing (this maneuver is called the *epoché*). According to Giorgi's (2009) descriptive phenomenology, the researcher must refrain from bringing in her own non-given past knowledge to account for whatever she is trying to understand. In fact, the researcher is to approach the research subject in a "naïve, pretheoretical way" (p. 135). This rigorous procedure entails bracketing ones scientific understandings, past knowledge, and personal experiences with a phenomenon. Perhaps more surprisingly, we also find this decidedly Husserlian strategy in Max van Manen's explicitly and emphatically hermeneutic phenomenology. When describing the epoché, van Manen (2014) talks about the mind being "cleared of garbage" (p. 223) and about "freeing oneself from obstacles" (p. 228). Some phenomenological researchers are so anxious to avoid distorting access to the phenomenon in question that they deliberately refrain from reading research literature until after carrying out their analyses (Ashworth, 1996). All these recommendations seem to rely on the idea that theoretical knowledge unduly restricts and narrows ones focus—Overgaard (2015) even calls it "the curse of knowledge."

I find it useful to dispel this restrictive view of knowledge. The idea that knowledge blinds us is simply a reversal of the old Platonic dogma that only by gaining knowledge do we see what *really* is. But "the reversal of a metaphysical statement remains a metaphysical statement," as Heidegger reminds us (2011, p. 158). Although keeping an open mind is surely an admirable ideal, I worry that dogmatically applying the epoché may lapse into willful ignorance. When it comes to research, I subscribe to the pragmatic point that knowledge shapes what we see and do as researchers. As a qualitative researcher, I do not pretend to let a phenomenon "show itself as itself" through some sort of mystifying epoché. When I interview students, for instance, I do not just sit back and let rich and detailed narratives flood me. I actively prompt, encourage, question, and respond to each remark. In this interaction, I neither remove nor bracket any knowledge. On the contrary, my entire research process is closely informed by my reading of postphenomenological theory, which I have used as a tool to help me generate empirical material. Empirical research is never a view from nowhere. Data are always codetermined by theory and methodology and in this respect, data is not simply given; it is *taken* (Brinkmann, 2014a). This point is to be understood in a positive way. As St. Pierre (2011) reminds us, "If we don't read the theoretical and philosophical literature, we have nothing much to think with during analysis except normalized discourses that seldom explain the way things are" (p. 614). Indeed, most mainstream psychologists would marvel at the idea of exploring human attention through anything other than rigorously designed randomized controlled trials.

Postphenomenological Validity

The argument that data is not simply given but taken raises the important question of validity: How do we assess the quality of analyses? According to van Manen (1990), a good phenomenological description *resonates* with lived life and evokes the phenomenological nod of recognition. The vital factor is not that phenomenological research claims correspond to an objective reality cleansed of human interest ("subjective bias"), but that they tap into a realm shared of experiences (Friesen, 2012). In a discussion of Iris Marion Young's feminist phenomenology, Ferguson (2009) likewise argues that a basic phenomenological test of validity is whether a description resonates with us, whether we see it as valid in our own experience: "We can only generalize from these experiences for those with whom they resonate" (p. 54). Applying resonance as validity criterion steers between the two extremes of false universalism ("everybody shares the same experiences") and idiosyncratic subjectivism ("each person's experiences are unique"). Using Young's article "Throwing Like a Girl" (1980) as example, Brinkmann (2012) similarly argues that phenomenology has the ability to "make the obvious obvious" by pointing toward something that may have remained unnoticed all along, but which we recognize when it is pointed out to us. The goal of phenomenology is not to shock or surprise, but to strike a chord of familiarity with its reader.

Validity judgments regarding the notion of a media portal thus ultimately lie with the readers of this chapter, but many technology users, this author included, can probably recognize the experience of experiential transportation through media technologies. Everything else withdraws and fades into the background. Even one's physical situatedness tends to recede. As a person that intensely dislikes flying, I frantically cling to such experiential transportation during transatlantic flights. Wellner (2014b) similarly states, "No wonder that dentists install televisions on the ceilings of their clinics so that the patients' attention is somewhat distracted from the teeth and the treatment" (p. 58). As a branch of social science that finds its strength in the intersubjective attunement implied by a notion of resonance, however, we must acknowledge that descriptions of common experiences may hide conflicts and indicate agreement where none is to be found (Friesen, 2012). In this regard, it is important to acknowledge that Wellner (2014a) mentions a gender divide in which male scholars seem to be cautious about multitasking, whereas female scholars tend to agree that it is possible. Indeed, the idea that women are superior multitaskers is popular in both conventional wisdom and the media. This is in line with Kent's remark that, "You always hear that men can't multitask." Against this idea, however, I cautiously note that Karen along with several of the female students I interviewed were quite skeptical of this gendered cliché (see also Aagaard, 2016b). Nevertheless, I would not claim to have the final word on this subject.

CONCLUSION: THE ETHICS OF ATTENTION

In conclusion, I want to relate the metaphor of a portal to the broader field of postphenomenology. The present discussion namely resembles another debate on media multitasking, which has arisen within the field of postphenomenology (see special issue of *Techné*; Wellner (Ed.), 2014). The core issue in this debate is using a cell phone (or "celling") while driving. On the one hand, Wellner (2014b) argues that we are developing a new attentional skill, *multi-attention*, which should be celebrated and practiced. Wellner (2014b) argues, "It is up to us to train and re-wire our brain in a manner that would support our beliefs, desires and ideologies" (p. 66). Hence, she views celling while driving in a favorable light. On the other hand, Rosenberger (2014) warns us against distraction caused by celling while driving and favors stricter regulation. How do we make sense of the fact that both authors employ postphenomenology, yet reach fundamentally opposing conclusions? I would argue that the answer hinges on the dual role of *possibility* in postphenomenology.

First, there is a distinction between *possible* and *actual*: While Wellner and Rosenberger agree on the postphenomenological notion of multistability, Wellner builds on the "multi-" part and imagines possible future variations, while Rosenberger stresses the "stability" of our present sedimentations. Historically, anti-essentialist strivings led Ihde to accentuate the "multi-" part of multistability by using variational theory to reveal additional gestalts (or stabilities) in visually ambiguous patterns, such as the Necker cube (Ihde, 1977). Later, this emphasis on variation and multiplicity was carried on to the analysis of technologies. As a result, postphenomenologists are generally eager to refute simplistic accounts of human-technology relations by showcasing alternative possibilities, a tactic known as the negative use of multistability (Rosenberger & Verbeek, 2015). In the cell-driving debate, for instance, Ihde (2014) has critiqued existing empirical studies for being rigid and narrow compared to phenomenological studies, because these studies focus on already sedimented actions in contrast to the phenomenological focus on possibilities. While this expansive "but wait, there's more!" use of multistability constitutes an excellent argument against the rigidity of technological determinism,

I worry that an over-emphasis of hypothetically possible human-technology relations may gloss over problematic actual relations and empirical patterns in our contemporary world. An overly optimist approach to multistability may ultimately render postphenomenology politically and ethically neutered. As an alternative, I would stress the importance of an empirically informed postphenomenology that departs from the complex and nuanced realities of our actual technological lifeworld and includes its downsides and drawbacks. As Foucault (1997) once put it, "My point is not that everything is bad, but that everything is dangerous, which is not exactly the same as bad. If everything is dangerous, then we always have something to do. So my position leads not to apathy but to a hyper- and pessimistic activism" (p. 256).

This leads me to a second distinction, namely between *possible* and *desirable*: It is undoubtedly true that we humans are fundamentally adaptive creatures that will adjust to most technological changes. Perhaps we can even adjust to a future world in which we concentrate only in short spells and resort to technologies at the slightest lull in our conversations or if a topic does not immediately interest us. The question, however, is not only whether such a future is possible, but also whether from a contemporary standpoint we deem such an existence desirable. While animals are guided by fixed appetites and desires, we humans have a capacity for so-called strong evaluations, that is, concerns with the normative worth of our desires: So even if it is in principle possible to develop multi-attention, why should we strive to do so? Which reasons (or "beliefs, desires, and ideologies") do we have for practicing multitasking instead of erring on the side of caution? Such questions concern the normative dimension of our media use, or *the ethics of attention*. Multitasking enthusiasts claim that driving while celling enriches the experience of driving so that it becomes more "vibrant" (Wellner, 2014b; Michelfelder, 2014). In other words, people should multitask in order to free themselves from encumbrances and boredom. But is minimization of discomfort an ethically strong argument? Attractive as the idea may be, there seems to be more to appraising the normative ramifications of multitasking than the joy of individual users: What about virtues like patience, trust, responsibility, and commitment? What if your vibrant experience happens at the expense of friends or fellow road users? Or, in other words, do I not owe my fellow human beings a certain attentiveness? There are good reasons to believe that I do. Shannon Vallor (2015) has recently argued that paying attention is not just a cognitive ability, but a moral skill:

> A person who cannot be counted on to pay attention when you tell her about the recent death of your closest friend, or who is unable to stay focused on the grave and imminent danger to which you're trying to alert her, or who cannot attend

> to the expressions on your face during an intense conversation, is not someone who can be said to be virtuous. (Vallor, 2015, p. 117)

In agreement with this sentiment, I caution against what I have elsewhere called *mediated impatience* (Aagaard, 2015).

Let me interject a final note about the advantages of discussing media portals. Framing the debate in terms of absence and presence sidesteps the objection that critics of multitasking tend to confuse attention with "full concentration." Wellner (2014b), for example, distinguishes everyday attention from full concentration and argues that multitasking may prohibit full concentration, but it does not prohibit splitting ones attention evenly between two tasks such as driving and celling. Since driving rarely requires our full concentration, she argues, multitasking is uncomplicated. Making a similar point, Diane Michelfelder (2014) argues, "If the opposite of distracted driving is driving while giving it one's full and complete attention, it would seem this is something we do not do on a regular basis" (p. 123). The need for full concentration, in other words, arises infrequently and only in specific situations. According to this chapter, however, what is problematic about multitasking is not so much a lack of full concentration as it is lack of *presence* in one's immediate circumstances. It is not that entering a media portal somehow prohibits an ordinarily concentrated being-in-the-world (a claim which, as both Wellner and Michelfelder rightly point out, would be phenomenologically untenable), but that it limits responsiveness to one's immediate surroundings due to its immersive nature. In social situations, this lack of responsiveness may be frustrating and deplorable to friends and conversational partners. In traffic situations, it may even be lethal.

REFERENCES

Aagaard, J. (2014). Media multitasking, attention, and distraction: A critical discussion. *Phenomenology and the Cognitive Sciences, 14*(4), 885–896.

———. (2015). Drawn to distraction: A qualitative study of off-task use of educational technology. *Computers & Education, 87*, 90–97.

———. (2016a). Breaking down barriers: The ambivalent nature of technologies in the classroom. *New Media & Society* (advance online publication). doi: 10.1177/1461444816631505

———. (2016b). Mobile devices, interaction, and distraction: A qualitative exploration of absent presence. *AI & Society, 31*(2), 223–231.

Ashworth, P. (1996). Presuppose nothing! The suspension of assumptions in phenomenological psychological methodology. *Journal of Phenomenological Psychology, 27*(1), 1–25.

Bennett, S., & Maton, K. (2010). Beyond the “Digital Natives” debate: Towards a more nuanced understanding of students’ technology experiences. *Journal of Computer Assisted Learning, 26*(5), 321–331.

Brinkmann, S. (2012). *Qualitative Inquiry in Everyday Life: Working with Everyday Life Materials*. London: Sage.

———. (2014a). Doing without data. *Qualitative Inquiry, 20*(6), 720–725.

———. (2014b). Psychiatric diagnoses as semiotic mediators: The case of ADHD. *Nordic Psychology, 66*(2), 121–134.

Brown, M. (2005). Learning spaces. In D. Oblinger & J. Oblinger (Eds.), *Educating the Net Generation*. Boulder, CO: EDUCAUSE.

Ferguson, M. (2009). Resonance and dissonance: The role of personal experience in Iris Marion Young’s feminist phenomenology. In A. Ferguson & M. Nagel (Eds.), *Dancing with Iris: The Philosophy of Iris Marion Young* (pp. 53–67). Oxford: Oxford University Press.

Foucault, M. (1997). On the genealogy of ethics: An overview of a work in progress. In P. Rabinow (Ed.), *Ethics, Subjectivity and Truth: The Essential Works of Michel Foucault 1954–84* (Vol. 1). New York, NY: The New Press.

Friesen, N. (2011). *The Place of the Classroom and the Space of the Screen: Relational Pedagogy and Internet Technology*. New York: Peter Lang.

———. (2012). Experiential evidence: I, we, you. In N. Friesen, C. Henriksson, & T. Saevi (Eds.), *Hermeneutic Phenomenology in Education: Method and Practice*. Rotterdam: Sense Publishers.

Giorgi, A. (2009). *The Descriptive Phenomenological Method in Psychology: A Modified Husserlian Approach*. Pittsburgh, PA: Duquesne University Press.

Hayles, K. (1999). *How we Became Posthuman*. Chicago, IL: University of Chicago Press.

Heidegger, M. (1995). *The Fundamental Concepts of Metaphysics: World, Finitude, Solitude* (W. McNeill & M. Walker, Trans.). Bloomington, IN: Indiana University Press.

———. (2011). A letter on humanism. In D.F. Krell (Ed.), *Basic Writings* (pp. 141–182). New York, NY: Routledge Classics.

Ihde, D. (1977). *Experimental Phenomenology: An Introduction*. Albany, NY: SUNY Press.

———. (2014). Embodiment and multi- versus mono-tasking in driving-celling. *Techné, 18*(1/2), 147–153.

Kvale, S., & Brinkmann, S. (2008). *InterViews: Learning the Craft of Qualitative Research Interviewing* (2nd ed.). Thousand Oaks, CA: Sage.

Lakoff, G., & Johnson, M. (1980). *Metaphors We Live by*. Chicago, IL: University of Chicago Press.

Mangen, A., & Kuiken, D. (2014). Lost in an iPad: Narrative engagement on paper and tablet. *Scientific Study of Literature, 4*(2), 150–177.

McCullough, M. (2013). *Ambient Commons: Attention in the Age of Embodied Information*. Cambridge, MA: MIT Press.

Merleau-Ponty, M. (1964). The philosopher and his shadow. In *Signs* (R.C. McCleary, Trans.) (pp. 159–181). Evanston, IL: Northwestern University Press.

Michelfelder, D. (2014). Driving while beagleated. *Techné, 18*(1/2), 117–132.
Overgaard, S. (2015). How to do things with brackets: The epoché explained. *Continental Philosophy Review, 48*(2), 179–195.
Pool, M., Koolstra, C., & van der Voort, T. (2003). The impact of background radio and television on high school students' homework performance. *Journal of Communication, 53*(1), 74–87.
Prensky, M. (2001). Digital Natives, Digital Immigrants, Part I. *On the Horizon, 9*(5), 2–6.
Radman, Z. (1997). *Metaphors: Figures of the Mind*. Dordrecht: Kluwer Academic Publishers.
Richardson, I., & Wilken, R. (2012). Parerga of the Third Screen: Mobile Media, Place, and Presence. In R. Wilken & G. Goggin (Eds.), *Mobile Technology and Place* (pp. 181–197). New York, NY: Routledge.
Rosenberger, R. (2014). The phenomenological case for stricter regulation of cell phones and driving. *Techné, 18*(1/2), 20–47.
Rosenberger, R., & Verbeek, P.-P. (2015). A field guide to postphenomenology. In R. Rosenberger & P.-P. Verbeek (Eds.), *Postphenomenological Investigations: Essays on Human-Technology Relations* (pp. 9–41). London: Lexington Books.
Schön, D. (1993). Generative metaphor: A perspective on problem-setting in social policy. In A. Ortonoy (Ed.), *Metaphor and Thought* (2nd ed.) (pp. 137–163). Cambridge: Cambridge University Press.
St. Pierre, E. (2011). Post qualitative research: The critique and the coming after. In N. Denzin & Y. Lincoln (Eds.), *The SAGE Handbook of Qualitative Research* (4th ed.). Thousand Oaks, CA: SAGE.
Tapscott, D. (1998). *Growing up Digital: The Rise of the Net Generation*. New York, NY: McGraw-Hill.
Vallor, S. (2015). Moral deskilling and upskilling in a new machine age: Reflections on the ambiguous future of character. *Philosophy & Technology, 28*(1), 107–124.
van Manen, M. (1990). *Researching Lived Experience: Human Science for an Action Sensitive Pedagogy*. Albany, NY: SUNY Series in the Philosophy of Education.
———. (2014). *Phenomenology of Practice: Meaning-giving Methods in Phenomenological Research and Writing*. Walnut Creek, CA: Left Coast Press.
Wellner, G. (2011). Wall-window-screen: How the cell phone mediates a worldview for us. *Humanities and Technology Review, 30*, 77–103.
———. (2014a). Celling while driving: Guest editor's introduction. *Techné, 18*(1/2), 3–9.
———. (2014b). Multi-attention and the horcrux logic: Justifications for talking on the cell phone while driving. *Techné, 18*(1/2), 48–73.
Wellner, G. (Ed.). (2014). Celling while driving [Special issue]. *Techné, 18*(1/2).
Young, I. M. (1980). Throwing like a girl: A phenomenology of feminine body comportment, motility and spatiality. *Human Studies, 3*(2), 137–156.

Part II

SELF-TRACKING AND IMAGING TECHNOLOGIES

Chapter 4

Human-Technology Relationships in the Digital Age

The Collapse of Metaphor in Biohacking

Moa Petersén

INTRODUCTION

In their (2012) book *Inventive Methods*, Celia Lury and Nina Wakeford argue that methodology hasn't developed in the same pace as the renewed interest for empirical and interdisciplinary studies within social and cultural sciences. Postphenomenology is among those theoretical frameworks where methodology lags behind. Postphenomenological methodology is still searching for a stable level of validity. As the digital culture and technology increasingly encapsulate our everyday lives, it is high time to find one. To quote Diane P. Michelfelder, "Now is a good time to be a postphenomenologist" (Michelfelder, 2015, p. 237).

The relationship between humans and technology within postphenomenology is a complicated one. The easy part is that there *always is* a relationship. There are no humans separated from technology and no technology separated from humanity—there are only hybrids. Postphenomenological theory on human-technology relations departs from four human technology relations observed and theorized by Don Ihde (1990): embodiment relation, alterity relation, hermeneutic relation, and background relation. Newly developed biotechnologies and VR environments demand new technology relations to be articulated, as they "fuse" with our biological body in a way not prior seen (Verbeek & Rosenberger, 2015, pp. 20–21). In my recent, ongoing, research I have set out to investigate biohacking groups as plausible new arenas for bio-knowledge production. In the following text, I will discuss how the human-technology relation I've found within the Swedish biohacking community becomes an important factor in explaining biohacking as new arena for civic science.

My investigation has so far shown that, in the biohacker community, the fusion between the biological body and technology seems to be rather radical: the biological body is believed *to actually be* a computer system, or better, a system of information. In my research project, I have chosen to analyze this fusion from a postphenomenological perspective; and in the following text, I will give examples of the methodological complexity of such a project. I have faced three main methodological problems that will be specifically addressed in this text. First, is it methodologically possible to use Ihde's technology relations in order to map the biohacker informants' relations to their biological bodies if the boundary between body and technology has been eradicated? Second, how can we move between a network level of the larger knowledge-producing community of biohacking, and the individual level where the biohackers' personal experiences are analyzed? Third, what is the validity of a study like this?

This text is based on my ongoing empirical research, and these three methodological questions will be discussed in relation to my account of my research process and results so far. In the first section, I open with a discussion of the possibilities a postphenomenological researcher has to transition from a network level to an individual level (and vice versa) in the same study. I then briefly contextualize biohacking—and also say something about self-hacking, which is a related phenomenon that is used occasionally in the text to further pinpoint my methodological challenges. In relation to this, I point out the human-technology relationship(s) I have found among the informants in terms of the "collapsed metaphor," and relate these to Ihde's embodiment relation, as well as to Ihde's theory on body one/body two (2002). This section is followed by some methodological reflections on validity of postphenomenological studies in general. After this, I explain and give examples of my efforts to find methodologically satisfactory postphemomenological solutions to carry out a study of the *maker approach* within new environments for bio-scientific production. By combining the concept of *biomedia* and the concept of *scientific literacy,* I suggest that we in biohacking groups—and perhaps in other new civic knowledge-producing environments as well—can see a biomedia transition from metaphorization to automatization, as learners are collectively engaging practically with the material in a way that does not presuppose, or even wish for, prior theoretical knowledge about explanatory models or their inbuilt metaphors. It is my overall hypothesis that a collapse of metaphor within these new science-producing environments boosts a "maker approach," as the stage of scholarly theoretical metaphorization is erased and the learners instead are led directly into the automatized stage that encourages activity and hands-on engagement with a material.

FROM BODILY SELF-UNDERSTANDING TO ACTIONS WITHIN THE NETWORK

This study takes place on two levels: when mapping the Swedish biohacking movement, I have studied rhetorics, discourses, and actions using a netnographic method. When moving to the individual perspective, I have made deep qualitative interviews with two members of the movement, aimed at identifying the informants' perceived relations to technology and their biological bodies.[1] By comparing the results from the interviews with Ihde's four technology relations, I have tried to postphenomenologically map what relations to technology are present among the movement's members. Two interviews is not sufficient empirical material for drawing confident and valid conclusions; this text should therefore be seen as the account of a pilot project, and the arguments put forward here should be treated as ideas rather than truths.

The transition back and forth between the level of the networked Swedish biohacker organization and the individual level of the biohackers' perceived personal beliefs toward their relation to their biological bodies and technology, made me undertake investigations into the possibilities of combining an Ihdean perspective with a more Latourian approach. The focus of my study lies at the formation of biohacking as a new arena for bio-scientific knowledge production, and the aimed result thus belongs more to a network level. Nevertheless, I believe postphenomenological theory and methodology which, in contrast to, for example, an ANT perspective, take the human body into account, provide an effective way to lay bare certain human-technology relations that I am convinced are shaping the overall methodology and scientific approach of the biohacking movement. Therefore, what my study needs is a marriage between an ANT perspective, through which actants and actions as parts of a larger scientific network can be studied, and a postphenomenological perspective that sees to the individual and existential bodily self-understanding. In my opinion, the level of bodily self-understanding has a high impact on the formation of the overarching biohacking system. In *What things do* (2005), Peter-Paul Verbeek shows how Don Ihde's and Bruno Latour's seemingly contradictory perspectives actually can be amalgamated. Verbeek argues that the very kernel of Ihde's and Latour's perspectives are the same—to overcome the subject-object distinction; and, comes to the conclusion that they can complement each other. The main difference that Verbeek sees is how the concept of mediation differs between the two perspectives. To Latour, a network is shaped by how the actants (that is hybrids of humans and their tools) within that network act. For Ihde, mediation theory deals with the experience of the world that the individual human has through or with technology. Verbeek argues that both thus overcome the distinction between object-subject, but in slightly different but coupled ways (Verbeek, 2005). Ihde sees the two theoretical systems as resting on "interrelational ontologies," but that the philosophical traditions where these ontologies come from differ. The more semiotic approach within ANT thus collides with the embodied phenomenology of postphenomenology (Ihde, 2015).

In my study, I have followed the path outlined by Verbeek as I argue that the two methods can complement each other. I have not tried to fuse them into each other, but rather let them complement each other from separate places. In summary, I believe it is crucial to study the actions (hands-on methodology) of the actants (the members of the biohacking movement) within the network (biohacking movement) in order to understand the formation (new arena of knowledge production) of this network. But, I argue, in order to understand the actions, it is also crucial to understand the experiences (perceived human-technology-body relations) of the individual members of the network.

CONTEXT: BIOHACKING

The biohacking movement was established in the United States in 2005. In summary, the biohacking movement conducts forms of genetic engineering outside of institutional settings. Experiments are varied but can include, for example, building lab equipment or analyzing human or animal genomes. Biohacking is a scattered phenomenon in the sense that it contains a lot of different subgroups with different types of research interests, goals, and ideologies. Sociologist Alessandro Delfanti has described biohackers as "life scientists whose practices exhibit a remix of cultures that update a more traditional science ethos with elements coming from hacking and free software" (Delfanti, 2013, p. 1). I have concentrated my research on the Swedish nonprofit biohacking organization BioNyfiken. BioNyfiken is the only biohacking organization in Sweden and was established in 2014. It consists of three subgroups located in the three main cities of Sweden: Stockholm, Göteborg, and Malmö. The three subgroups together have approximately 750 members (as of November 2015). Within BioNyfiken Sweden, research is focused on making RFID/NFC implants into the human body, by manipulating the brain through electrical impulses to optimize its performance, and by hacking bacteria and DNA. The methods of engaging with the biological material are very hands-on. The implants, the brain manipulations, and the DNA-hacking are all described as being possible to carry out using household equipment. For instance, the brain manipulation machine (tDCS) is built with a Wettex dish cloth, tinfoil, an old mouse pad, and hot melt adhesives.

BioNyfiken Sweden is partly integrated with Quantified Self Stockholm, a nonprofit organization for self-trackers (564 members in November 2015). Also, the Quantified Self movement is of US origin, and was founded in 2007 by *Wired* magazine editors Gary Wolf and Kevin Kelly. This community, as such, is not central to my study, but in this methodological reflection I will use it occasionally in order to further explain the technology relation present within the biohacking community. In the self-hacking movement, so-called Quantified Self-technologies (QS-technologies)—such as health applications, sleep trackers, and fitness trackers—are used to extract and review biological body data in order to manipulate and optimize the bodily organism or its "output" and "performance." One of the main concerns that Quantified Self Stockholm and BioNyfiken Sweden share is optimization of personal and public health. It is believed that body-hacks will lead to increased individual and public health through heightened individual responsibility. The strong aim of life improvement often found within biohacker and self-hacker groups is an ethos appropriated from computer hacking environments (Levi, 1984/2010; Delfanti, 2013, pp. 12–13). As part of this hacker ethos, both communities contest the system where large established institutions are in charge of research on the human biological body and medicine. Biohacking and self-tracking both point to their possibilities of solving problems that institutionalized research has hitherto failed to solve.

When blog posts and home pages are analyzed, an explicit analogy of human biological body and computer(-system) comes forth in the rhetorics of both movements. The human body is conceptualized as possible to hack, optimize, and control. This is not true only for the Swedish examples, but serves as a general belief also for the movements on an international scale. On the homepage of one of the largest self-hacker movements in the world, Self-hacker.com, for example, we can read, "As an individual, you are your own architect or software engineer that need to program your body based on your requirements." Likewise, my interview material shows that informants from both movements tend to describe their biological bodies as computerized systems, or systems of information. BioNyfiken operates on a more societal level, as their main goal is to contest the established bio-research system by providing the general public with a new scientific environment outside of the bio-research industry and academia. The members of Quantified Self Stockholm, on the other hand, are more interested in personal bodily optimization. Thus it would be fair to propose that BioNyfiken is more into hacking the scientific system and the general human biological system, while Quantified Self Stockholm is more concentrated on hacking the personal bodily system to optimize bodily factors, such as personal strength, health, and endurance.

THE COLLAPSED METAPHOR

I started the analysis by looking at the verbal use of the body/computer(-system) analogy that I had found both within the organizational rhetorics and the personal statements. Since Ihde's theory on human-technology relations neither includes thoughts on the "pure" human biological body (compare body 1/body 2) or body-technology verbal metaphorization, I turned to theories on the metaphor in order to open up the material collected. George Lakoff and Mark Johnson (1986) have famously argued how metaphors shape actions. But how this action is shaped in reality is relatively under-explored and under-theorized (van der Weele & van den Boomen, 2008). In this project I argue that something I call "the implosion of the metaphor" encourages action within the biohacker environment.

In the case of BioNyfiken, the relation to technology seems to have moved inside the body and merged with the relation to the bodily biological self. How can we explain this from a postphenomenological view? First some clarifications. The conceptualization of the biological human body as a computer system which is possible to "hack," control, or optimize, present within the biohacker movement, can be interpreted from many different perspectives. The most obvious would perhaps be as part of a long-lived cultural futurist trope of the cyborg (i.e., Haraway, 1991) that in more recent years—fueled by fast-developing inventions on the biomedical consumer market and of wearable technology alike—has fused with a transhumanist ideology. Further on, technological metaphors for the human body are by no means a novelty. Both the scientific explanatory models of the biological body, as well as the popular concept of the body, have a long history of mirroring technological development, and new technological innovations (Coleman & Fraser, 2011). Within biological science, explanatory models to the biological body that contain metaphors from the digital sphere—such as code and information—have a long history (Kay, 2000; Keller, 2002). Eugene Thacker has researched the metaphorical development within biology induced by the formation of bioinformatics during the 1980s, where data science and biological science amalgamated to an extent not prior seen (Thacker, 2004, p. 33). Without investigating it further Evelyn Fox Keller notices how both the border between organisms and computers, and the border between computer science and organism science, gradually have been made more and more porous as an effect of developments of computer programs and computer visualization techniques that have narrowed the epistemological gap between sample and its carried data (Keller, 2002, p. 203). Thacker continues this observation and takes it one step further as he argues that within the new digitally framed biology science, not only has the border been made porous, but it has in fact collapsed into one and the same leveled dimension of biomedia. This is much due to biology's accelerating dependency on and use of computer technology (Thacker, 2004, pp. 39–40). Thacker analyzes the example of "information" that has been a metaphor for DNA since the 1950s, and argues that information no longer can be called a metaphor for DNA, but instead must be viewed as an inherent part of DNA as a technical principle (2004, p. 40). The relation between information and DNA, for example, has thus been transitioned "from metaphorization to automatization" (Thacker, 2004, p. 40), in a process where no external appropriation of metaphors is needed anymore. Thus, biomedia is a step away from the classical sense of a metaphor where it is described as "a figure of speech in which a word or phrase is applied to an object or action to which it is not literally applicable" (Oxford Dictionaries).

It's not hard to find support for a biomedia collapse of the metaphor in contemporary biological research. As an example, a group at Zurich's Department of Chemistry and Applied Biosciences has recently showed that contemporary alternatives of storing data on memory sticks and hard drives are outclassed by new methods of storing data on DNA. Compared to the 50 years that data lasts on a CD, storing data in bone tissue will make it last for a million years, at least (Grass et al., 2015). Body, DNA, and information are inseparable in this example. This interlace is clearly illustrated by the image that was made by one of the researchers in the group to visually enhance the research results (Figure 4.1). In the illustration, the data digits overlap seamlessly with the genetic information within the bone piece, as if they naturally belonged to the same system of information.[2]

Another example of a complete merge between body, DNA, information, and technology is the recent developments of biological engineering within synthetic biology. Here a total merge between DNA code and computer code is commonly agreed upon as the base for research. On the blog *bionyfiken.se*,[3] Sina Armoor Pour, one of the co-founders of BioNyfiken Sweden,

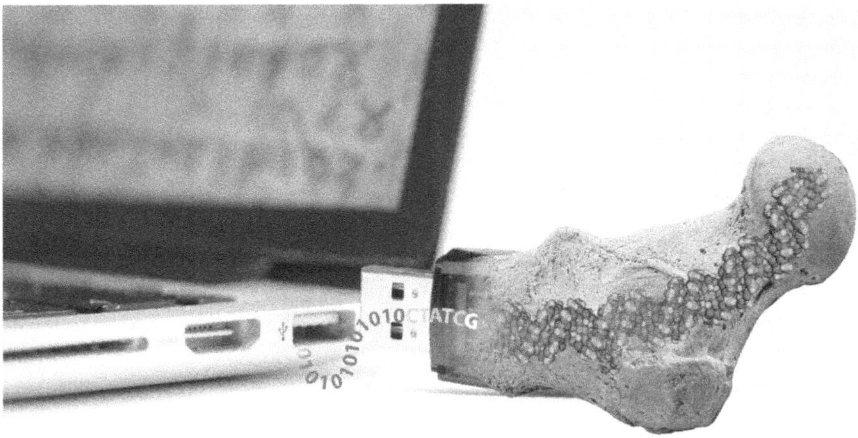

Figure 4.1 Illustration (Philipp Stössel, ETH/Zurich). Previously published in Grass et al., 2015. Used with permission. (Philipp Stössel from Grass, R. N.; Heckel, R.; Puddu, M.; Paunescu, D.; Stark, W. J., Angew. Chem.-Int. Edit. 2015, 54 (8), 2552–2555. Reprinted with Permission.)

answers a question what BioNyfiken will do in their newly built laboratory in Stockholm. He answers by referring to a TED talk by bioengineer Tal Danino where he presents his aims to develop techniques for programming bacteria with different algorithms, as is done in computer software. The desired outcome of Danino's project is to program these bacteria to detect and treat diseases like cancer. The example clearly shows a transition from metaphorization to automatization that it shares with a lot of contemporary research within synthetic biology, genetic engineering, and biomedicine. In Danino's example, bacteria will not only be understood as digital devices, they will also be treated practically as digital tools with different functions that are determined by the programmers' individual intentions. Armoor Pour answers that this is the kind of research BioNyfiken wants to engage in. This proves how the collapse of the metaphor is also of high importance to the Swedish biohacking movement. Another post on the *bionyfiken.se* blog (April 25, 2015), in which Armoor Pour reflects upon an article on biohacking that had been published the previous day, confirms that the collapse of metaphor is important to BioNyfiken's approach to bio-research. In the article, published in Sweden's largest daily newspaper *Dagens Nyheter*, the journalist explains the methods of a biohacker by comparing them to the methods of a computer hacker, someone who "rebuilds, experiments, and creates new within computer technics. The biohacker wants to do the same, but with organic material instead of PCBs and code" (Larsson, 2015). Armoor Pour reacts to this by explaining that biohackers who work with bacteria and DNA instead "use DNA and bacterias as computer hackers use computers, PCBs, and code. That's where we are heading. A future where we use organic machines *as* computers and PCBs and where we write code with DNA."[4] The journalist uses a metaphorical explanation where the biohacker is supposed to do something to the organic material with inspiration from a model coming from outside of the biological field. In Armoor Pour's answer, this metaphor is collapsed and computer and body are made equivalent to each other in an amalgamated "organic machine" actually consisting of PCBs and code.

Now it is time to look at the relations that surfaced in the interviews in comparison to Ihde's four technology relations, in order to find how to methodologically proceed from here. First, let me briefly sketch Ihde's human-technology relationships (2015) to assure my departing points:

1. *Embodied relation*—Ihde's classical example is looking through a pair of glasses. Even if the glasses are not explicitly noticed all the time, it co-shapes our relation, engagement, and interactions with our surrounding world.
2. *Alterity relation*—In this relation we treat our technology at hand as if it was a living creature separated from us.
3. *Hermeneutic relation*—Ihde's classic example to exemplify this relation is the thermometer that we have to "read" and interpret in order to turn into a perception. We will not actually experience heat or cold, but the interpretation of what the thermometer displays will give us a representation of reality that we can translate into our own bodily sensation of temperature.
4. *Background relations*—in this relation the technology makes up the background of our experience, and creates a context for our perceptions.

After interviewing the biohacker informants, the material retrieved showed that the human-technology relation that was dominating among the biohackers was most equal to the embodied relation. The seamless blending of the biological body and the computer system within the movement shows affinity with Ihde's example with the glasses that grow onto the bearer and become an unconscious extension of the biological body. The transparent way in which it shapes the world- and life-experience of the glass bearer is also obvious in the biohacker example as ideas about the computer system leak onto the conceptualization of the human biological body, and also shape the actions taken to manipulate, optimize, or control it. But the blending of body and technology is not in itself an object for reflection among the biohackers. The border between biological body and computer system in the biohacker environment has imploded in such a way that it avoids reflection and conscious meditation. Just as with all of Ihde's relations, the embodied relation in Ihde's scheme is dependent on intentionality; and the relation between the glass bearer and the glasses never frees the glasses from being a tool used by the human individual in order to enhance reality. Our intentionality to engage with the world is mediated through the tools. Thus, there is always a possibility to take the glasses off and stop the enhancement. The human relation to technology as a tool cannot be entirely unconscious, except, perhaps, for short time periods. Moreover, I'm sure the level to which "transparent" embodiment of technological artifacts is possible is a strictly individual matter. The amount of factors that impact how prone a certain individual is to reaching successful transparent embodiment is probably vast, and is—though it is not within the scope of this text—an interesting and much needed field for future postphenomenologists to map. In summary, within the biohacking community, both intentionality and relation have been reduced to a minimum, as the tool dimension—to a large extent—has been eradicated.

To make this a bit clearer we can compare it to the relation I found present among the self-hacker informants. Though the rhetorics on a network level seemed to show a parallel implosion of metaphor present both within the biohacking community and the self-hacking community (optmizing and controlling the biological body as a computer), interviews laid bare that different kind of technology relations dominated the two groups. From the interviews with the self-hacker informants, it was also hard to distinguish between the informants' relations to their technological equipment and their relations to their own biological bodies. The equipment they used to monitor biodata was perceived of as displays making the data—already present within the body—visible. Just as in the example with the USB-bone piece (Figure 4.1; Grass et al., 2015), the data code of the technological equipment and the biological data were seamlessly blending into another, and flowing from body to device without friction. The body was further understood as a computer system consisting of data and information that was partly controllable but also unpredictable in its network structure. The informants' approaches to their own biological bodies made me think of the rhizomatic network described by Deleuze and Guattari (1980/1987). One of the informants expressed how he experienced how this partial feeling of control made him insecure, and that this came from ideas of the body working as an open-ended web or network. He explained his fears that the disruption or manipulation of one node of the network within the system may cause changes (often long-term) that would eventually lead to illness and instability of the system. This comprehension of the biological body shows affinity with a rhizomatic network model that is decentralized and unpredictable; and one that finds new ways of transferring information even if it has been manipulated in order not to (introduction; Deleuze & Guattari, 1980/1987). Moreover, among the self-hackers, I found something that resembled a hermeneutic relation, where the informants reported they were interpreting the signs of the body parallel to the information that their technological devices visually displayed. One informant also reported how her technological devices led her to questioning if her experience of her own body was, in fact, correct. For instance, she reported how the devices made her think and react in ways like, "My blood pressure has risen—I must be feeling bad." The technological devices thus led her to feel bodily sensations she would not have felt if she hadn't been able to interpret information on the displays. She reported this as "anxiety rising." The hermeneutic relation, in this case, thus lead to something that can be compared to Ihde's alterity relation. I have elsewhere related the alterity relation to Sigmund Freud's concept of the uncanny, as feelings of quasi-otherness and experiences of the uncanny, in my opinion, share a profound likeness (Goysdotter, 2013). The experience of the body turning into someone to which I necessarily relate but don't really trust shows signs of a failed embodiment relation. The embodiment process has then stopped halfway and, due to different factors, avoided implosion. This could be compared to the relation found among the biohackers where the separation between body and technology found within the self-hacker movement has been overbuilt and no glitch to reflect or hesitate in front of it is present. If the self-hackers could be placed closer to the hermeneutic/alterity relation not reaching the embodiment relation, the biohackers have transgressed the embodiment relation and instead moved toward a collapse of the relation.

Recent postphenomenological studies have dealt with contexts of new technology where Ihde's four human-technology relations have been somewhat twisted toward what Peter-Paul Veerbeek and Robert Rosenberger call a "fusion relation" (Verbeek & Rosenberger, 2015, p. 21). In these relations, found, for example, in cochlear implants or artificial heart valves, the relation is more intimate than that active in the embodiment relation (p. 21). I suggest that the collapse of the metaphor rather suggests that the relation between human biological body and technology within biohacking has imploded into a *non-relation*. It seems to me as if a belief in that the body actually *is* a computer could hardly be called a human-technology relation anymore. For what is a relation, in a phenomenological understanding, if there is no conscious distance between the components? The non-relation is closely knit to Ihde's theory on the interlaced body one and body two (2002). According to Ihde, body one is the perceiving, active, here-located, bodily experience developed by Maurice Merleau-Ponty as the *corps vécu*. Body two is the cultural or socially constructed body, discussed by Michel Foucault, for example (Ihde, 2002, p. 69ff.). It seems as if the biomedia collapse of technology into the human body produces a body concept that is neither possible to explain by body one *nor* by body two. Both the concept of body one and of body two are depending on a relation to the surrounding world—either as a world of things to perceive as is the case of body one, or as a world of social constructions as in body two. The body concept that surfaces through the collapsed metaphor, I will suggest we call *body zero.* This body is not a perceiving or a relating body, but a body which it is possible to do things to and engage practically with. Body zero is of course interlaced with body one and two, and every human contains all of them. But, I suggest, body zero could be studied on its own in order to reveal, for example, the hacker approach to the human biological system found within biohacking. In the next paragraph I will develop this idea further.

THE MAKER APPROACH POSTPHENOMENOLOGICALLY UNDERSTOOD

At the network level, we saw how biohackers and self-hackers seemed to share the same relation to technology as dissimilated with their biological bodies. But a closer phenomenological dive into the conceptualizations of the individual members of the movements showed how the human-technology relations differed between the two communities. Such a zooming-in that detects differences can be related to Ihde's concept of *multistability.* Multistability is an interesting concept when discussing validity of postphenomenological studies, as it can, in fact, pose a threat against the validity of postphenomenological methodology. According to the multistability principle, we can never approach technology as if it was a homogenous entity. Technology must instead always be treated as *technologies*, an approach that avoids us from falling into the trap where single observations are taken to be great truths. Into this principle goes also that technologies never are to be considered autonomous, that they are non-neutral and therefore ambiguous, and that their effect on life-experience differs between cultures (Ihde, 1990, p. 144). It is clear in the case of self-hacking and biohacking that Ihde's multistability principle was proven right. At a more particular level, in combination with a phenomenological methodology, the stability that was first observed showed to be multistable. But where do we reach a level where this eternal instability of the research areas and the empirical material can be solid enough to draw a valid conclusion? How can a research method studying relations that continually hybridize avoid being caught in the same unstable and non-valid state? This is, of course, not a methodological question only relevant for postphenomenological studies, but is a classic critique against, for example, social constructivist studies. So, what criteria for validity are there for postphenomenological studies? The principle of multistability puts postphenomenology in a place from where it is risky to draw general conclusions. Sure, to draw the conclusion that technologies are multistable is a sort of conclusion in itself, but what should we do if we want to be able to speak about the network in which the postphenomenologically approached individuals are active? In my efforts to make my study gain a higher level of validity, I have tried out a cyclic process. That is, I have started at the network level and moved towards the individual level. In this process I have destabilized (or multistabilized) the network level, and then turned back again to the network level. Let us now, lastly, see what happens if we follow this cyclic process. Thus, the next question of my study: How can we take further the human-technology non-relation present within biohacking, to shed light on the movement's approach to science-making, if the movement is seen as a new arena of bio-scientific knowledge production?

Biohacking groups are new environments for scientific production and can be examined against the concepts of civic participation, civic science, and citizen science (i.e., Jasanoff, 2005, 2012). As such, biohacking could be studied as part of the open science movement, which is an important factor in the recent struggle over redefinitions and reshaping of scientific institutions and public participation in scientific knowledge production (Delfanti, 2013, pp. 10–11). These new environments have in the last years been strongly emboldened as digital channels for tutorials, teaching, and research has increased (Silvertown, 2009). *Scientific literacy* is an important concept within the discussion of new arenas for production of scientific knowledge. It is a huge discussion, and I will only briefly mention it here since, in my case, it serves to explain the methodological differences between the biohacking movement and institutionalized bio-science. The traditional definition of scientific literacy is one that relies heavily on the individual understanding of scientific concepts and methodology. National Science Education Standard defines scientific literacy as "the knowledge and understanding of scientific concepts and processes required for personal decision making" and as "the ability to describe, explain, and predict natural phenomena," (National Academy of Sciences, 1996, p. 22). Scientific literacy is also used by the biohackers, but in their use it has taken on a very different shape. In "A Biopunk Manifesto"[5]—which is the main document of the biohacker movement, written by one of its central figures Meredith Patterson—scientific literacy is held forth as "necessary for a functioning society in the modern age. Scientific literacy is not science education. A person educated in science can understand science; a scientifically literate person can 'do' science" (Patterson, 2010). When presenting the manifesto at the UCLA conference Outlaw Biology? Public Participation in the Age of Big Bio (2010), Patterson herself illustrated the biopunk focus on doing in favor of understanding, in the introduction to her talk, as she showed a chart displaying all of the human metabolic pathways. When showing it, Patterson said, "When I look at that, I think 'Wow, this is cool, what can I DO with it?'" (Patterson, 2010); this opposed to another possible formulation that might have been, *what can I understand from it*?

I suggest that the concept of "scientific literacy" as it is used within the biohacker movement is directly connected to the collapse of the metaphor and the non-relation to technology; and that they, in turn, are constitutive for the biohacking movement as environment for learning. I have earlier in this text discussed a quote from Amoor Pour where he defines the methodology of biohackers when they engage with their material. Above, I used it to highlight the imploding metaphor found in biohacker groups, but it could also be used to illustrate how an implosion of metaphor is a methodological step from imitating, understanding, and metaphorization to automatized doing, using, and crafting. In the biohacker environments, as in the computer hacker environments, the practice of the members is described as highly material. Firstly, they meet in so-called Makerspaces. Moreover, biohackers often describe themselves as carpenters, tinkerers, or other craftsman-oriented professions. As argued in the former paragraph, the meaning of scientific literacy, to the biohacker movement, is to know what one is able to do with something on a material level when presented with it—just as a carpenter knows what could be done given a piece of wood and a chisel. This practical outlook is inseparable from the biohacker view on the human biological body as programmable, hackable, and controllable.

In the Makerspace, learners are collectively engaging practically with the material in a way that does not presuppose, or even wish for, prior theoretical knowledge about explanatory models or their inbuilt metaphors. Rather it is seen as a relief if no homogenous pre-understanding is present among the participants. Further on, a heterogeneity of the group composition is desired in order to reach innovative solutions. Co-founder Hannes Sjöblad tells in an interview, found on the BioNyfiken homepage, that "the Swedish biohacker community is a very diverse group of people with different backgrounds and interests: experienced lab researchers, quantified self-enthusiasts, body modifications artists, chip implantees, makers, grinders, and fitness- and performance optimizers. This is a bubbling scene, when these groups come together in the same room, incredibly creative interactions happen." Amoor Pour expresses a similar idea as he writes about his wishes for a nearby future where "people without any biological education whatsoever use the technology in order to investigate themselves and their environment and push the biotechnological development forward." The BioNyfiken environment is thus explicitly structured to bring non-specialists together to act upon things together—not, in the first hand, to understand these things together. This environment seems to fulfill the meaning of scientific literacy as it has been explained by Roth and Calabrese Barton, who argue that scientific literacy is something that is gained through an interactive situational exchange between individuals, organizations, and subjects of study through multiple forms of discourse practices (Roth & Barton, 2004, p. 105). BioNyfiken also identify themselves as in opposition to the institutionalized research environments and to the methods applied in these environments that Sjöblad, in the interview referred to above, calls "a closed sector only accessible to people with specialized training wearing white lab coats." Resistance of this kind is characteristic to hacker environments, and is interlaced with the understanding of scientific literacy found in the biohacker environment. So, I argue, the focus on doing instead of understanding within the new environment of knowledge production of biohacking groups has its roots in the members' biological self-understanding as computer systems. The barriers between model and real object have been dissimilated on the body-conceptualization level, and this is what leads the actions of the researchers in their doer-approach to the human biological body.

CONCLUSION

In this text, I have focused on three postphenomenological methodological problems of my present investigation into the Swedish biohacking community. The first focus has been on how to make the transition between the network level and the individual level. In this study, this transition is of great importance since I argue that the conceptualization of the body at one level connects to the conceptualization of knowledge production, or methodology, in the other—and vice versa. The conceptualization of the biological body as a hackable system, and the hands-on methodology that I have found within the Swedish biohacker community, seem to corroborate that such a transition is possible and valid. The other focus has been on how to apply Ihde's human-technology relations to a situation where technology has merged totally with the body in a biomedial collapse. The relation to the body within biohacking seems to parallel best Ihde's embodiment relation, but it calls for a new and even more intimate relation (I have called it a non-relation in this text). Even if Ihde's four relations can't sufficiently explain what has happened in biohacking, the validity of using his human-technology relations on this material becomes evident when compared and contrasted to the relation present within self-hacking where the same bodily conceptualization was present on the network level. The use of Ihde's postphenomenological human-technology relations here has shown a discrepancy on the individual level that wouldn't have been detected if only the network level would have been studied. The third focus has been directed toward how to increase the validity of a postphenomenological study like this. One answer, I suggest, is to combine a study of the network level with a postphenomenological study of the individual level—or a systematic analysis with phenomenological methodology—in a cyclic process. In such a process, a seemingly stable network context can be deconstructed by postphenomenological multistability, and the multistability could be slowed down and controlled by the overarching systematic context.

NOTES

1. With the same aim, I have also made deep interviews with two members of the quantified self-movement.
2. The image was published together with the article (Grass et al., 2015).
3. http://bionyfiken.se/
4. All translations from Swedish are my own. Italics are my emphasis.
5. "Biopunk" is an umbrella term for many different sorts of biohackers with different foci. A biohacker is thus a biopunk, but a biopunk is not necessarily a biohacker.

REFERENCES

Coleman, D., & Fraser, H. (2011). Introduction—minds, bodies, machines. In D. Coleman & H. Fraser (Eds.), *Minds, Bodies, Machines, 1770–1930* (pp. 1–11). Basingstoke: Palgrave Macmillan.

Deleuze, G., & Guattari, F. (1987). *A Thousand Plateaus: Capitalism and Schizophrenia*. London: Bloomsbury Academic. (Originally published 1980)

Delfanti, A. (2013). *The Politics of Open Science*. London: Pluto Press.

Goysdotter, M. (2013). *Impure Vision: American Staged Art Photography from the 1970s*. Lund: Nordic Academic Press.

Grass, R. N., Heckel, R., Puddu, M., Paunescu, D., & Stark, W. J. (2015). Robust chemical preservation of digital information on DNA in silica with error-correcting codes. *Angewandte Chemie International Edition*, *54*(8), 2552–2555.

Haraway, D. (1991). A cyborg manifesto: Science, technology, and socialist-feminism in the late twentieth century. In D. Haraway (Ed.), *Simians, Cyborgs, and Women* (pp. 149–181). New York, NY: Routledge.

Ihde, D. (1990). *Technology and the Lifeworld: From Garden to Earth*. Bloomington, IN: Indiana University Press.

———. (2002). *Bodies in Technology*. Minneapolis, MN: University of Minnesota Press.

———. (2015). Positioning postphenomenology. In R. Rosenberger & P.-P. Verbeek (Eds.), *Postphenomenological Investigations. Essays on Human-Technology Relations* (pp. vii–xvi). London: Lexington Books.

Jasanoff, S. (2005). *Designs on Nature. Science and Democracy in Europe and the United States*. Princeton, NJ: Princeton University Press.

———. (2012). *Science and Public Reason*. Oxon: Routledge.

Kay, L. (2000). *Who Wrote the Book of Life? A History of the Genetic Code.* Palo Alto, CA: Stanford University Press.

Keller, E. F. (2002). *Making Sense of Life. Explaining Biological Development with Models, Metaphors, and Machines.* Cambridge, MA & London: Harvard University Press.

Lakoff G., & Johnson, M. (2003). *Metaphors We Live By.* Chicago, IL & London: University of Chicago Press. (Originally published 1980)

Larsson, L. (2015, April 24). De hackar livets byggstenar. *Dagens Nyheter.*

Levi, S. (2010). *Hackers: The Heroes of the Computer Revolution.* Sebastopol, CA: O'Reilly Media. (Originally published 1984)

Lury, C., & Wakeford, N. (Eds.). (2012). *Inventive Methods: The Happening of the Social.* London: Routledge.

Michelfelder, D. P. (2015). Postphenomenology with an eye to the future. In R. Rosenberger & P.-P. Verbeek (Eds.), *Postphenomenological Investigations. Essays on Human-Technology Relations* (pp. 237–246). London: Lexington Books.

National Academy of Sciences. (1996). *National Science Education Standards.* Washington DC: National Academies Press.

Patterson, M. (2010). A biopunk manifesto [public lecture]. Delivered at *Outlaw Biology? Public Participation in the Age of Big Bio, 8th Annual Symposium.* UCLA Center for Society and Genetics. Retrieved from https://vimeo.com/18201825.

Roth, W. M., & Barton, A. M. C. (2004). *Rethinking Scientific Literacy.* New York, NY: Routledge.

Rosenberger, R., & Verbeek, P.-P. (2015). Introduction. In R. Rosenberger & P.-P. Verbeek (Eds.), *Postphenomenological Investigations. Essays on Human-Technology Relations* (pp. 1–6). London: Lexington Books.

Silvertown, J. (2009). A new dawn for citizen science. *Trends in Ecology & Evolution, 24*(9).

Thacker, E. (2004). *Biomedia.* Minneapolis, MN: University of Minnesota Press.

van der Weele, C., & van den Boomen, M. (2008). How to Do Things with Metaphor?: Introduction to the Issue. *Configurations, 1*(16), 1–10.

Verbeek, P.-P. (2005). *What Things Do. Philosophical Reflections on Technology, Agency, and Design.* University Park, PA: Pennsylvania State University Press.

Chapter 5

Service Interfaces in Human-Technology Relations

A Case Study of Self-Tracking Technologies

Fernando Secomandi

Postphenomenology is regularly lauded by philosophers of technology for providing impetus to a recent "empirical turn" in the field (Achterhuis, 2001; Ihde, 2009; Rosenberger & Verbeek, 2015; Verbeek, 2005). From a philosophical perspective, to be empirical, among other things, means distancing oneself from sweeping analyses at high abstraction levels, in order to focus on the particularities of concrete practices of technology design and use. Postphenomenologists hold that by such an approach one was better positioned to avoid undesirable transcendentalist, dystopian, and nostalgic tendencies that characterized much of classical philosophy of technology.

For those familiarized with research methods in the social sciences, on the other hand, to be empirical means generating first-hand data through interviews, observations, collection of documents, and so on, often by spending significant time "in the field." In this sense of the word, very few postphenomenological studies may be called empirical (e.g., Aagaard, 2014; Forss, 2011; Friis, 2015; Hasse, 2008; Secomandi, 2013), for the majority of publications depend on syntheses of secondary bibliographic sources. The point is not made as a criticism but simply as an observation that those who want to conduct postphenomenological field studies must look elsewhere for methodological guidance. And in the process of doing so, they can end up exploring phenomena from novel perspectives, arriving at new findings for postphenomenological theory.

This chapter is an exemplification of that—an empirical investigation of a self-tracking technology, inspired by ethnographic approaches to the study of service design practices, resulting in the acknowledgment of an overlooked topic in postphenomenology: the intersubjective constitution of human-technology relations. The self-tracking technology in question is DirectLife, a service commercialized by Philips to assist users in becoming more physically active (http://www.directlife.philips.com).[1]

Also known as activity or fitness trackers, self-tracking technologies are systems that provide users insight into activity patterns, typically with the goal of supporting behavioral change and improving general well-being. These technologies often consist of a wearable electronic device that employs biometric sensors plus networking technologies to record users' behavioral or physiological data and sync these with another device (e.g., smartphones) for further data crunching and visual feedback.

In the case of Philips DirectLife, the wearable component is an activity monitor supposed to be worn by users on a daily basis in the trousers' pocket, as a necklace, and occasionally in the socks when cycling. By plugging it into a personal computer, the monitor automatically uploads activity data to the DirectLife website application on which dashboard users can track their recorded levels of physical activity. In the DirectLife website, users also follow a 12-week activity improvement program consisting of progressively higher targets to be reached on a weekly basis.

Many self-tracking technologies like DirectLife are embedded in programs formulated according to "sports training" or "gaming" logics that are geared at enhancing users' commitments to activity targets. These programs may include virtual communities built around users' existing social networks or managed by manufacturers, where people can compete among themselves and/or share information and tips. At the time of its introduction, DirectLife was distinctive for offering users one-on-one supervision from professional coaches via e-mail communications. Because of the rich data that can be generated about people's physical and social activities, as well as the potential that is created by Internet-based connectivity, self-tracking technologies hold for companies the promises of better customizing offers for different users and groups, of building stronger and more profitable relationships, and gaining strategic knowledge about target markets.

The topic of self-tracking technologies has attracted considerable attention in academic circles, not least by philosophers of technology (see van den Eede, 2015, for a review). This chapter is an extension of studies that adopt a postphenomenological standpoint (Irwin, 2016; Secomandi, 2013; van den Eede, 2015). In what follows, DirectLife is addressed as a *service interface*. What this term suggests is the position that technologies can assume in mediating exchanges between the people who exchange a particular service (Secomandi, 2013; Secomandi & Snelders, 2011, 2013). By focusing on the service exchange relations that embed a self-tracking technology like DirectLife, one can discover how human-technology relations are formed intersubjectively by users and providers. Specifically in this study, the uncovering of the interpersonal dimension of technological experience is partly made possible by a research methodology that highlighted the social dynamics regarding the design and use of the DirectLife service.

The next section starts with an explanation of how self-tracking technologies may be studied as service interfaces from a postphenomenological perspective. The following section, then, elaborates upon the topic of intersubjectivity as a dimension of human-technology relations. These conceptual sections prepare for the empirical study of DirectLife, which describes how experiences of this self-tracking technology are constituted intersubjectively by users and designers interacting through service interfaces. The chapter concludes with a discussion about intersubjectivity in relation to postphenomenology's interrelational framework.

SERVICES, POSTPHENOMENOLOGY, AND SELF-TRACKING TECHNOLOGIES

Services have long been deemed peculiar economic entities because of the closer forms of human interaction that are required for the production of outcomes. Researchers from diverse disciplines, such as economics, marketing, operations, geography, and industrial design largely work under the shared assumption that, for the production of services to occur, users must contribute information, labor, possessions, or even their own bodies, as inputs to provider operations. In short, services are cocreated by providers and users. Yet, the meaning of this assertion can only be grasped in contrast to how goods have constantly been portrayed following what economic theorists known as the "anonymity principle" (Gallouj & Weinstein, 1997, p. 540). The principle holds that upon production a good typically attains a degree of physical independence from those who produced it and those who will consume it, which allows it to circulate freely in the economy beyond the original context of production. As shown in the history of economic thought, the belief that services lack the material independence and permanence of goods was operative already three hundred years ago, when Adam Smith famously declared unproductive the labor of domestic servants by explaining that it

> does not fix or realize itself in any particular subject or vendable commodity. His services generally perish in the very instant of their performance, and seldom leave any trace of value behind them, for which an equal quantity of service could afterwards be procured. (cited in Delaunay & Gadrey, 1992, p. 12)

The productivity of service activities is thus negated because they are embodied in a form that is not physically bound, storable, or tradable, as goods purportedly are. Nowadays, service research is a long way from the view above that services are simply immaterial and unproductive, yet the question of how they differ from goods is not completely resolved (see Illeris, 2007).

It is not necessary to fully enter the goods versus services debate to remark that, from a postphenomenological perspective, one can immediately object to the portrayal of goods as "anonymous" entities. A brief description of the fountain pen like the one used to draft this chapter would suffice to problematize such a characterization. In use, the pen is in a position of mediation between paper and writer. It transforms expressive intentions into scribed text, in ways that would be impossible without it. Here is a first challenge to the independence of the pen—what it is and does can only be determined by taking into account a practice of writing that entangles a human user.

A second challenge surfaces when noting how writing, although predominantly located in the private domain of a study desk, is not entirely solitary. As the user attends to the brand name embossed on the pen's body or the concavities near its tip affording a firm grip, a relation with its designers is established. To be sure, in ordinary writing, the felt presence of these "other" humans is most negligible, so the tendency is to ignore them. Few would describe the experience of writing with a pen mainly in terms of an exchange relation with its designer. However, under rigorous postphenomenological analysis, the dimension of human interactivity might be recoverable even in the simplest case of a mass-produced good.

Contrast this example with a visit to the doctor—a classic service case—and the interpersonal relation needed to cocreate an outcome becomes prominent. In such situations it is difficult to dissociate the "service" from the human interactions giving rise to it, even if what is exchanged between doctor and patient also involve partly "objectified" bodily aspects, such as physical attributes, outward appearance, embodied skills, expressed emotions, and so forth. To conclude, experiences with fountain pens and medical consultations are differently graded in terms of human-to-human interactivity.

In service research, Shostack (1977, 1982, 1987) was perhaps the first to describe services from an experiential perspective. With a background in marketing and operations, she conceived of services as essentially intangible processes that are organized by providers in order to deliver benefits to consumer. Because delivery processes were in principle "hidden inside" organizations, consumers could only experience services through their "peripheral clues," or "evidence," which Shostack (1977, p. 77) defined as "the things that the consumer can comprehend with his five senses—tangible things."

In the case of air transportation, for example, the tangible evidence would include flight tickets, onboard food and drinks, the airplane itself, even flight attendants. Through the service evidence, users could indirectly perceive and evaluate "backoffice" provider-controlled processes, for instance, by forming an impression of what went on inside the plane's cockpit when hearing the pilot over the loudspeaker.

The term "evidence" might be highly suggestive for phenomenological ears, but Shostack's construal of service experience was in fact markedly unphenomenological. The tangible evidence was conceived as a surrogate "reality" for users, whereas the "essence" of a service was to be found in intangible processes over which only service providers could gain direct knowledge. In Secomandi and Snelders (2011), a critique of Shostack and revision of her two-domain account of services is advanced on basis of the concepts of *service interface* and *service infrastructure*, the former representing socio-technical resources that are closely associated with user-provider exchanges, while the latter represents resources that are indirectly related to user-provider exchanges and need to be actualized through an interface. From the embodied standpoint of users or providers, these two domains are entwined in such a way that service *reality* can only be experienced in interaction with material interfaces and the infrastructural aspects that are enacted through them.

In postphenomenological terms, it is possible to further elaborate the interplay between service interfaces and infrastructures through an analysis of *technological mediation* (Secomandi, 2013; Secomandi & Snelders, 2013). By applying Ihde's framework of human-technology-world relations to services, the experiences of users or providers is thus formalized in terms of *human-interface-infrastructure* relations, whereby the other human who cocreates the service experience is revealed in technologically mediated encounters as an aspect of the infrastructure. Explored from the perspective of users, this framework clarifies that service experiences can display the different modalities of technological mediation first identified by Ihde (i.e., embodiment, hermeneutic, alterity, and background). This line of research focusing on user experiences is expanded by Secomandi (2013) through an empirical study of Philips DirectLife, the same self-tracking technology that is analyzed in the present chapter.

Despite the new terminology of service interface and infrastructure, the application of postphenomenology to the study of self-tracking technologies is not an uncharted terrain. Broadly speaking, postphenomenological literature is filled with detailed case descriptions of user experiences with screen-based, networked, or wearable electronic devices, that share relevant characteristics with self-tracking technologies (e.g., Aagaard, 2014;

Bakardjieva, 2005; Ihde, 2007, 2010; Irwin, 2016; Rosenberger, 2009, 2013b; Verbeek, 2008; Wellner, 2015). Fewer studies, however, investigate how technology can mediate human experience of another human being (or bodily aspects thereof) (Forss, 2011; Rosenberger, 2011b; Verbeek, 2008). In particular, little has been written about technologically mediated experience of a person's *own corporeality*. Ihde (2002) deals at times with issues of virtual embodiment in gaming and virtual reality, as well as of computer-generated medical imagery of his own body (Ihde, 2009). Besmer (2015) describes tele-operated robotic arms as an extended avatar of an operator's body. Forss (2011) briefly comments on occasion in cancer cell screening through microscopes where cytodiagnosticians analyze their own material. What is common throughout these cases is the identification of one's own body as the primary aspect of the world that is transformed in a technologically mediated situation. For self-tracking technologies, this is key to their capabilities to track the *self*.

At this point in the discussion I wish to raise two questions in respect to how self-tracking technologies can be studied as services from a postphenomenological standpoint. First, there is the question of how exactly users' bodies are revealed for them through a service interface. This reflexive characteristic of self-tracking technologies was problematized recently by van den Eede (2015, p. 153), who wondered how user bodies may be postphenomenologically accounted for as "worldly" aspects: "When one considers graphics of sleep patterns, step count, distance travelled, et cetera, of what exactly are those images?" To be sure, with this question van den Eede alludes to two distinct but interrelated issues. On the one hand, there is a *reference* issue, of how one can experience what is depicted in these graphics as one's own corporeality. On the other hand, there is a *translation* issue, of how lived experiences of embodiment can be "read" from the graphical elements that are displayed onscreen. While this second issue is not relevant to the argument being developed in this text,[2] the first is, and in my understanding the trouble identified by van den Eede follows from the limited treatment of the concepts of *body schema* and *body image* in postphenomenological research (Besmer, 2015; Ihde, 2002; Secomandi, 2015; Welton, 2006). These terms represent a duality of human embodiment commonly described as the phenomenological difference between "being" and "having" a body. I hold that the main aspect of the world to which a self-tracking technology refers is the *body image* of a user. The empirical study of DirectLife below further complicates matters by showing how these body images can be revealed through service interfaces, and as such become part of the infrastructures that are partly controlled by service providers.

The second question that can be raised when examining self-tracking technologies as service interfaces regards which standpoint to adopt. The postphenomenological studies mentioned so far opt exclusively for the user perspective. In the study below, in addition to that I contemplate the provider perspective, more specifically, the experience of designers who implemented a new functionality on the DirectLife website. This take of the empirical research approximates it to ethnographic studies about the visualization practices of engineering designers in the field of science and technology studies (e.g., Bucciarelli, 1994; Henderson, 1999; Vinck, 2003). I highlight the social context of design work regarding the DirectLife service. I also argue that design visualizations can be approached as service interfaces with users and that designer experiences of these visualizations depend on how users are enacted through them. In conclusion, experiences of DirectLife, for both designers and users, are constituted intersubjectively on basis of their relations through a service interface. To support this assertion, a postphenomenological account of the role of intersubjectivity in human-technology relations is needed, a concern to which I turn next.

INTERSUBJECTIVITY IN HUMAN-TECHNOLOGY RELATIONS

Postphenomenological research has by and large overlooked the role of interpersonal exchange in human-technology relations. A notable exception is Forss's (2011) ethnographic study in a cytology laboratory, which contains fascinating detail of how expert evaluation of cell material through the microscope can depend on knowledge about the patients concerned. For example, knowing that a patient was exceptionally old could make cytodiagnosticians extra cautious in detecting cancerous cells that might lead to invasive surgery. Also, when checking samples from a patient one had met in person, diagnosticians could abstain from carrying out assessment and hand over the job to colleagues.

Other exceptions in postphenomenological literature are the studies that address the topic of human-to-human contact through telecommunication technologies (Bakardjieva, 2005; Ihde, 1990, p. 199; Rosenberger, 2011b; Wellner, 2016). These latter studies establish a valuable counterpoint to the argumentation I seek to develop in this section, because they offer an opportunity to ask whether any technologically mediated encounter between human beings implies the exchange of services.

In special, Bakardjieva's (2005) integration of sociological phenomenology and critical theory with Ihde's (1979) analysis of the amplification-reduction structure of technologies is the most attuned to the issue of intersubjectivity. Bakardjieva observes that telecommunication technologies, in general, have amplification powers to enlarge one's social world, by providing knowledge about previously unknown people and societies, by connecting geographically distant individuals, by fostering new relationships, and so forth. On the other hand, the type of social encounters that happen through these technologies can be constrained in terms of tangibility, reciprocity, intimacy, and so on. Of special concern is the fact that these technological reductions can be "designed in" telecommunication technologies by its producers and thereby shaped according to a capitalist logic of socioeconomic exploitation. The danger would be that such a logic then frames every tele-encounter made possible by the technology, resulting in the alienation of its users.[3] About the technological reductions caused by the telephone, Bakardjieva concludes: "Thus, I have to express my love for my distant mother through the sole medium of my voice amplified by the telephone/telecommunication system, over a period of time constricted by the high cost of the long distance call" (Bakardjieva, 2005, p. 65).

The passage, however, does not sufficiently detail how producer logic is felt as operative within the technologically mediated experience. When someone speaks to a distant mother over the phone, one also finds himself relating to the mobile network provider. Differently from the mother's voice, to which focal attention is directed, the network provider is not the immediate interlocutor. Its presence is experienced rather more indirectly, for instance, as the growing feeling of time scarcity that contextualizes the conversation that unfolds. It is from its field position that the network provider bestows significance upon how mother appears to the caller. She may seem indifferent to the rising costs of the call, as seemingly insignificant moments in the day are recounted in minute detail. Reflexively, the caller may become aware of his own impatience, monetary stringency, or even rudeness.

This analysis reinforces Bakardjieva's point by clarifying how the socioeconomic logic of service exchange with a service provider can subtly encroach upon social relations through a technological system that the provider partly controls. To be sure, Bakardjieva herself does not believe domination by the "System" to be the necessary outcome of technological mediation. She goes on to elaborate a phenomenological conception of *home* as a personal realm where users are able to creatively appropriate technologies and renegotiate its mediation effects, thus resisting external control. At this juncture, however, my argumentation bifurcates from Bakardjieva's, for I believe that opportunities for user empowerment cannot be found in a domain that is totally devoid of technological influence.

Two tenets of postphenomenological philosophy of technology is that human activity is, and has always been, technologically embedded and that humans do not need to, and possibly cannot, be "outside" a technologically mediated situation in order to discern its non-neutrality (Ihde, 1990). In the example above, it is precisely because the background presence of the service provider can be discerned by the caller, that he is able to acknowledge, accept, or resist its influences when talking to other people over the phone. This is not the same as holding that interpersonal phone exchanges can happen outside the influence of network providers. The point is that influences from these providers should not be feared on basis of a misguided belief that they are entirely surreptitious and necessarily totalizing.

A few concepts from postphenomenological literature can be suggested for expanding the notion of intersubjectivity delineated above. Drawing on the Foucauldian concept of *self-practice*, Dorrestijn (2012) and Verbeek (2011) argue that people can adopt and rehearse particular techniques of use in order to take a stance toward the mediating effects of technologies. In the case of the phoning a family member, for instance, one could deal with the "reduction" caused by network-controlled technology by saving money for making long-distance calls, or by finding a calmer moment in a busy schedule in order to place them. Another useful concept is that of "hermeneutic strategy" (Rosenberger, 2011a, 2013a), which originally refers to perceptual habits and theoretical underpinnings that underline image interpretation in science. If the concept can be transplanted to the context of telecommunications, it might be useful to point to particular cultural predispositions that can shape mediated social encounters. For instance, one could imagine mother-daughter relations as forged upon a logic of caring and love that is, if not impervious, at least highly accommodating to the influences of the network provider.

These are relevant reminders that when engaging technologies human subjects are always corporeally and culturally situated. Ihde (1990, pp. 42–71) once posited that Galileo's scientific discoveries made possible by the telescope could only come about at a particular historical moment where special ways of experiencing space and time through technologies had already become sedimented in human perception. This cultural-historical embedding of vision is further investigated by postphenomenologists in a number of case studies about imaging technologies in science (e.g., Friis, 2015b, 2015a; Hasse, 2008; Ihde, 1998; Rosenberger, 2008, 2011a).

To the extent that the situatedness of human perception mentioned above can be in part attributed to sociality, postphenomenological research already addresses the issue of intersubjectivity. But the studies above seem off the mark of a true analysis of technological mediation. And that is because the explanatory effort is mainly directed at social interactions that transcend human-technology relations as such, in the sense of occurring "before" or "around" them. For instance, the interpretation of visual evidence can be explained with reference to previous training on the part of scientists or to their affiliation to different research communities. But an account of the intersubjective constitution of technologically mediated experiences should not be limited to explanations of how a particular perceptive act is affected by extraneously formed cultural-historical sediments. Instead, as in the case of telecommunications analysis above, it is necessary to demonstrate how social relations are enacted and are able to transform perception *from within* the technologically mediated encounter. The study presented next attempts to do precisely that, by describing how user and designer experiences are constituted intersubjectively in mediated relations through the service interfaces of DirectLife.

EXPERIENCING DIRECTLIFE'S SERVICE INTERFACES

Methodological Considerations

Data collection for this study of DirectLife ran from April to July of 2009, around the time when this service was launched in Dutch and North American markets. At that time, DirectLife was under responsibility of New Wellness Solutions (NWS), a new venture located at the Philips Incubator, in Eindhoven. They were experimenting with the business-to-business marketing strategy to sell DirectLife to organizations that would in turn make the service available for employees, in an effort to promote workplace health.

Part of the study strategy involved field research at the DirectLife development location, including observation of daily operations, in-depth interviews with members of the administrative and design staff, plus collection of documentation from the digital platform used by the software development team. These data concentrated on the design and implementation of a new functionality of the DirectLife website dashboard for users: the goal adjustment interface. The purpose of this interface was to allow DirectLife users to individually adjust their targets for the end of the 12-week activity improvement program on basis of information about actual performances during the initial weeks. The intent of the design team was to make users more committed to weekly targets, thus closer to reaching the end of the program successfully. The analysis below of the designer experience of a DirectLife service interface is based on data collected in this field study.

The other part of the research strategy involved a usability study of DirectLife. Besides completing a 12-week activity improvement program myself, six participants from a Dutch university were recruited to test the service for several weeks, until the day when they were offered the opportunity to adjust their activity targets using the new goal adjustment interface. At the end of their participation, semi-structured interviews were conducted with all users in order to delve into their impressions and experiences about DirectLife. This part of the study informs the analysis below about the user experience of the website interface.

The way in which this empirical research was planned and carried out had important implications for the findings that can be drawn from it. Assessing the current state of postphenomenological literature, when the topic of methodology is mentioned at all, it is mainly discussed in terms of broad research heuristics, rather than systematic tools and techniques. Recently, Rosenberger and Verbeek (2015, pp. 30–32) summarized the characteristics of postphenomenological methodology as (1) "understanding the roles that technologies play in the relations between humans and world," (2) "[including] empirical work as a basis for philosophical reflection," (3) "[investigating] how, in the relations that arise around a technology, a specific 'world' is constituted, as well as a specific 'subject'," and, as a result of the foregoing, and (4) "[making] a conceptual analysis of the implications of technologies." Ihde, in the same vein, characterizes postphenomenological methodology as the application of a Husserl-style "variational method" (e.g., Ihde, 1993, 2003, 2006, 2009); but whereas the method is carefully worked out through an examination of visual illusions in the book *Experimental Phenomenology* (Ihde, 1986), its application to the study of concrete technologies is much less rigorously specified (e.g., Ihde, 2012, in special, Part III). In a reflective note, Ihde (2003) has wondered if his own approach to study technologies, although motivated by simplicity and clarity, had a special "selectivity" for individual, subjective, and simple cases. To that, it would be possible to add that a great portion of published postphenomenological studies rely on secondary bibliographic sources, not primary data garnered by researchers' immersion in social settings.

With regard to my own selectivity, I took an alternate track than Ihde, because conceiving of DirectLife as a "service" technology focused the inquiry not on any individual perspective, but the exchange relations between providers and users. Even if I became a DirectLife user, conclusions could not be drawn solely from my own experience, but had to be juxtaposed with the impressions and interpretations of other participants of the usability study. Moreover, the choice to visit the DirectLife development site foregrounded the topic of interpersonal relations, for I became immersed in the social environment where real-life design work is done. With respect to this ethnographically inspired part, methodological support was found in a stream of science and technology studies about engineering design practices that displays a higher sensitivity to the social context of image interpretation and the data collection techniques involved in ethnographic research (e.g., Bucciarelli, 1994; Henderson, 1999; Vinck, 2003). Such a research strategy decisively inclined me into collecting data about and reflecting upon the intersubjective constitution designer and user experiences of DirectLife.

The User Experience of a DirectLife Service Interface

Figure 5.1 displays the *history view interface*, one of the main features of the DirectLife website for users, regularly visited more than once a day from the start of the activity improvement program (see Secomandi, 2013, for an analysis of user experiences with DirectLife involving other interfaces besides the website). In the history view, users primarily consult the graphics displayed at the center of the screen in order to gain knowledge about measured levels of physical activity happening in the previous hours, days, weeks, and months in the DirectLife program.

Users never experience the DirectLife history view independently from the influence of the organization behind it. To the extent that Philips is generally regarded a reputable innovator in the field of consumer electronics, the history view might be initially accepted as a true representation

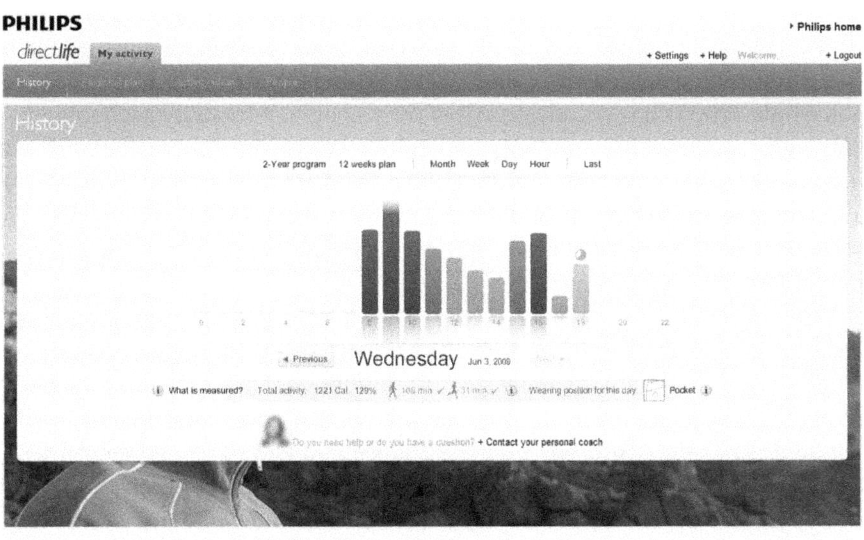

Figure 5.1 **DirectLife's history view interface (Fernando Secomandi).**

of users' bodily activities. However, incongruences are often noted between what is seen onscreen and what is believed to have actually elapsed in lived experience. One of the main sources of incongruence is related to how the activity monitor is worn. Users typically experiment with DirectLife recommendations on how to carry the activity monitor, for instance, wearing it inside a sock when cycling. Most come to recognize that the history view is not a "transparent window" into their bodily activities, but a reading of it performed by the monitoring technology that the DirectLife service employs.

This influence of the service provider can be felt even more strongly when considering the interactivity of the DirectLife website. Because the history view is accessed as an online application, a dialogue between service users and providers ensues. At the very least, users understand that what is displayed onscreen depends on instant retrieval, processing, and transmission of personal data on the part of DirectLife. Depending on users' technical proficiency, visualizations of physical activity through the history view can be textured by knowledge about dedicated algorithms to estimate calorie expenditure and calculate targets, for instance. Other, design-conscious users might notice in the visual elements of history view the "technical style" intended by its web designers, or other possible navigation options that were or were not made available by them.

Obviously, the most direct indications that DirectLife users interact with a service provider come from the personal coach image right below the history view and the Philips logo at the top-left corner. But while users can deliberately look and make sense of these peripheral visual elements, that is not what normally occurs as the history view is used. Much like the network provider in the case of telephone communication before, personal coaches and the Philips organization are at the fringes of experience, whereas the activity graphs of the history view focalize the attention of users. Nonetheless, these service providers influence the experience of users, in the case of coaches, by making history view readings potentially a motive to look for help, and in the case of Philips, by imbuing the whole experience with a sense of proprietorship and authenticity.

To be sure, the whole perceptual field of users is impregnated with the logic of their service exchanges with DirectLife and thereby arranged according to "regions" of particular significance. The appearance of the history view graphs at the very center communicates the kernel of DirectLife's value proposition: whether one is physically active or not. Peripheral options, including clarifications over contractual agreements and privacy policies, are secondary to that.

Postphenomenologically, I hold that the graphics of DirectLife's history view refers to objectified aspects of users' own bodies, more precisely, their body images. These body images are actualized through a DirectLife interface, and as such, they become an infrastructural aspect of the service experience of users. Because DirectLife has a role to play in these experiences, users' knowledge about their bodies are constituted intersubjectively in relation to the service provider.

The Designer Experience of a DirectLife Service Interface

Figure 5.2 displays one of the many visualizations that were generated during the development and implementation a new functionality to the DirectLife website: the goal adjustment interface (see Secomandi, 2012, pp. 81–118, for an analysis of the other visualizations used during the design of DirectLife). This particular visualization emerged in the process of "debugging" software codes of the goal adjustment interface, before they were merged into the DirectLife website. Testing happened on the "demo server," which held copies of website software and user data, in order to prevent disordering the software version that was online for users. Visualizations created in the demo server were based on physical activity data from actual users that were merged with other "bogus" information, like fake names, to ensure their anonymity. Software testers imagined a number of use cases for DirectLife users and rehearsed their behaviors by manipulating the goal adjustment interface, interpreting

Figure 5.2 DirectLife's goal adjustment interface (Fernando Secomandi).

what appeared onscreen on basis of their expectations of how users might understand that.

The visualization above shows one of the bugs found by a software tester. Very low activity counts are displayed for the previous several weeks in the goal adjustment interface, as can be seen in the bars referring to "assessment" and "weeks 1-3" (perhaps, the user has regularly forgotten to carry the activity monitor). The unmatched activity targets, the gray line partly covered by the "contact your coach" pop-up, nonetheless remain high and coherent with what is considered in this case a "healthy" level of physical activity. As the user reaches week 4 ("now") and is offered the opportunity to revise activity targets for the end of the 12-week improvement program, the goal adjustment toggle ("new goal") illogically locks his options at very low levels of weekly activity ("20 Cal"). DirectLife's promise to help people become more physically active would be seriously undermined if users were forced to set activity targets at such levels.

For software testers, this bug could be logically explained in terms of a poor fit between the new interface and underlying mathematical algorithms and software codes. These had to be rewritten. However, it is important to observe that the flaws were not discovered by directly examining algorithms and codes, but instead by noting an unexpected onscreen appearance. What's even more important is that this mismatch had to be experienced by the software tester vicariously, by putting himself "in the shoes" of future users and imagining what they would feel were the new interface implemented in the DirectLife website without further corrective actions.

From a postphenomenological standpoint, as in the case of the history view, the graphics displayed in the goal adjustment interface refer to body images. Differently from before, the objectified body is not an aspect of the own body of the designer who is experienc*ing* the interface, but of the user who is experienc*ed* through it. Actualized by the interface, the bodies of users are perceived by designers as infrastructural aspects of the service experience. Because this infrastructure entails information and behaviors that are attributed to the users of DirectLife, designer experiences of the interface are constituted intersubjectively in relation to them.

CONCLUSION

The chapter has developed several lines of reasoning that need integration in this conclusion. The underlying motivation was to remark that postphenomenological philosophy of technology can fruitfully adopt qualitative research methodologies from the social sciences. Doing so not only can help to "operationalize" postphenomenology when carrying out empirical research in social settings, the results from such undertakings can also bring original insights to postphenomenological theory. As mentioned before, the conclusions drawn from this study of DirectLife are partly due to the methodological decision to study "service" technologies following an ethnographic approach to design practices. This has presented an inclination to critically consider the socioeconomic exchanges that happen through a service interface, both from the user and the designer perspectives, and to delve deeper into the issue of intersubjectivity.

Specifically with regard to the case of self-tracking technologies, intersubjectivity follows from the observation that user bodies are being "tracked" through service interfaces, what makes the experience of them dependent on reciprocal exchanges with the providers of these services. That raises important questions about the "servicing" of bodies that are made possible by self-tracking technologies: If user embodiment is revealed through a service interface that is partly controlled by providers, then what is the nature and extent of this influence? What type of user subjectivity can be forged in mediated service encounters with providers of self-tracking technologies? These are relevant questions that might be asked from a postphenomenological standpoint, in line with its predominant interest on user perspectives. However, with this study I hope to have shown that the same questions can be aptly taken up from the standpoint of designers as well.

More generally, the contribution of this chapter can be related to postphenomenology's interrelational ontology. Postphenomenological interrelationality is itself an application of phenomenology's concept of intentionality to the case of human-technology relations (Ihde, 2003). Intersubjectivity, as discussed here, can be seen as a special dimension of human-technology relations referring to the significant ways in which mediated experiences can be textured by social interactions—in this case, interpersonal service exchanges. To explain such texturing postphenomenologists need not transcend a technologically mediated encounter in order to search "outside" or "backward" for the cultural-historical conditionings of human perception. Instead, a rigorous analysis of intersubjectivity can describe how social interactions are enacted and are transformative *in* the mediated experience.

By advancing the concept of service interface, I hope to make postphenomenological research better attuned to the issues of intersubjectivity. This, I believe, is important not just for understanding the case of self-tracking technologies, but the growing number of technologies that are exchanged in the marketplace as services, in areas such as education, health care, transportation, retail, tourism, and others.

NOTES

1. As of September 30, 2016, when this chapter was being finalized, the formerly official DirectLife URL redirected to a webpage where Philips communicated the discontinuation of the DirectLife service and website starting on July 1, 2016.

2. This problem refers to the distinction often made in postphenomenology between *isomorphic* and *translational* images (see Ihde, 1998). Van den Eede and I (Secomandi, 2013) apparently agree that the graphics generated by self-tracking technologies are mostly translational depictions of users' bodies, but he further complicates this point by observing how these visualizations can also be isomorphic. When seeing the numeric step count displayed by a pedometer, for example, the "feeling" one has can resonate (isomorphically) with the experience of having taken exactly that many steps. In fact, Ihde (1998, pp. 167–168) makes similar observations regarding a "vestigial analogue" referentiality of some translational visualizations, for instance, in the case of a thermometer's mercury column "embodying the 'higher' and 'lower' temperatures shown."

3. Adding slightly to Bakardjieva's point, there is no reason why designed-in mediations regulated by producers should only be worrisome for their technological reductions. As seen in the case of gambling practices, technologies can also "lock-in" users in exploitative relations with producers precisely because of their amplification qualities.

REFERENCES

Aagaard, J. (2014). Media multitasking, attention, and distraction: A critical discussion. *Phenomenology and the Cognitive Sciences*, 1–12. https://doi.org/10.1007/s11097-014-9375-x

Achterhuis, H. (Ed.). (2001). *American Philosophy of Technology: The Empirical Turn*. Bloomington, IN: Indiana University Press.

Bakardjieva, M. (2005). *Internet Society: The Internet in Everyday Life*. London: Sage Publications Ltd.

Besmer, K. M. (2015). What robotic re-embodiment reveals about virtual re-embodiment. In R. Rosenberger & P.-P. Verbeek (Eds.), *Postphenomenological Investigations: Essays on Human-Technology Relations* (pp. 55–71). Lanham, MD: Lexington Books.

Bucciarelli, L. L. (1994). *Designing Engineers*. Cambridge, MA: MIT Press.

Delaunay, J. C., & Gadrey, J. (1992). *Services in Economic Thought: Three Centuries of Debate* (1st ed.). Dordrecht: Springer.

Dorrestijn, S. (2012). Technical mediation and subjectivation: Tracing and extending Foucault's philosophy of technology. *Philosophy & Technology*, *25*(2), 221–241. https://doi.org/10.1007/s13347-011-0057-0

Forss, A. (2011). Cells and the (imaginary) patient: The multistable practitioner–technology–cell interface in the cytology laboratory. *Medicine, Health Care and Philosophy*, *15*(3), 295–308. https://doi.org/10.1007/s11019–011-9325-0

Friis, J. K. B. O. (2015a). Gestalt descriptions embodiments and medical image interpretation. *AI & Society*, 1–10. https://doi.org/10.1007/s00146-015-0615-6

———. (2015b). Towards a hermeneutics of unveiling. In R. Rosenberger & P.-P. Verbeek (Eds.), *Postphenomenological Investigations: Essays on Human-Technology Relations* (pp. 215–225). Lanham, MD: Lexington Books.

Gallouj, F., & Weinstein, O. (1997). Innovation in services. *Research Policy*, *26*, 537–556.

Hasse, C. (2008). Postphenomenology: Learning cultural perception in science. *Human Studies*, *31*(1), 43–61.

Henderson, K. (1999). *On Line and on Paper: Visual Representations, Visual Culture, and Computer Graphics in Design Engineering*. Cambridge, MA: The MIT Press.

Ihde, D. (1979). *Technics and Praxis*. Dordrecht: Reidel.

———. (1986). *Experimental Phenomenology: An Introduction*. Albany, NY: State University of New York Press.

———. (1990). *Technology and the Lifeworld: From Garden to Earth*. Bloomington, IN: Indiana University Press.

———. (1993). *Postphenomenology: Essays in the Postmodern Context*. Evanston, IL: Northwestern University Press.

———. (1998). *Expanding Hermeneutics: Visualism in Science*. Evanston, IL: Northwestern University Press.

———. (2002). *Bodies in Technology*. Minneapolis, MN: University of Minnesota Press.

———. (2003). If phenomenology is an albatross, is post-phenomenology possible? In D. Ihde & E. Selinger (Eds.), *Chasing Technoscience: Matrix for Materiality* (pp. 131–144). Bloomington, IN: Indiana University Press.

———. (2006). Forty years in the wilderness. In E. Selinger (Ed.), *Postphenomenology: A Critical Companion to Ihde* (pp. 267–290). Albany, NY: State University of New York Press.

———. (2007). *Listening and Voice: Phenomenologies of Sound* (2nd ed.). Albany, NY: State University of New York Press.

———. (2009). *Postphenomenology and Technoscience: The Peking University Lectures*. Albany, NY: State University of New York Press.

———. (2010). *Embodied Technics*. Copenhagen: Automatic Press/VIP.

———. (2012). *Experimental Phenomenology: Multistabilities* (2nd ed.). Albany, NY: State University of New York Press.

Illeris, S. (2007). The nature of services. In J. R. Bryson & P. W. Daniels (Eds.), *The Handbook of Service Industries* (pp. 19–33). Cheltenham: Edward Elgar Publishing.

Irwin, S. O. (2016). *Digital Media: Human-Technology Connection*. Lanham, MD: Lexington Books.

Rosenberger, R. (2008). Perceiving other planets: Bodily experience, interpretation, and the Mars Orbiter camera. *Human Studies*, *31*(1), 63–75.

———. (2009). The sudden experience of the computer. *AI & Society*, *24*(2), 173–180. https://doi.org/10.1007/s00146-009–0190–9

———. (2011a). A case study in the applied philosophy of imaging: The synaptic vesicle debate. *Science, Technology & Human Values*, *36*(1), 6–32.

———. (2011b). Embodied technology and the dangers of using the phone while driving. *Phenomenology and the Cognitive Sciences*, *11*(1), 79–94. https://doi.org/10.1007/s11097-011-9230-2

———. (2013a). Mediating Mars: Perceptual experience and scientific imaging technologies. *Foundations of Science*, *18*(1), 75–91. https://doi.org/10.1007/s10699–012-9286-7

———. (2013b). The importance of generalized bodily habits for a future world of ubiquitous computing. *AI & Society*, *28*(3), 289–296. https://doi.org/10.1007/s00146-012-0410-6

Rosenberger, R., & Verbeek, P.-P. (2015). A field guide to postphenomenology. In R. Rosenberger & P.-P. Verbeek (Eds.), *Postphenomenological Investigations: Essays on Human-Technology Relations* (pp. 9–41). Lanham, MD: Lexington Books.

Secomandi, F. (2012). *Interface Matters: Postphenomenological Perspectives on Service Design* (Doctoral thesis). Delft University of Technology, Delft.

———. (2013). Thinking through the service interface: A study of Philips DirectLife. *Design Philosophy Papers*, *11*(1), 65–88.

———. (2015). Bodies as technology: How can postphenomenologist deal with the matter of human technique. In R. Rosenberger & P.-P. Verbeek (Eds.), *Postphenomenological Investigations: Essays on Human-Technology Relations* (pp. 105–122). Lanham, MD: Lexington Books.

Secomandi, F., & Snelders, D. (2011). The object of service design. *Design Issues*, *27*(3), 20–34. https://doi.org/i: 10.1162/DESI_a_00088

———. (2013). Interface design in services: A postphenomenological approach. *Design Issues*, *29*(1), 3–13. https://doi.org/10.1162/DESI_a_00192

Shostack, G. L. (1977). Breaking free from product marketing. *Journal of Marketing*, *41*(2), 73–80.

———. (1982). How to design a service. *European Journal of Marketing*, *16*(1), 49–63.

———. (1987). Service positioning through structural change. *Journal of Marketing*, *51*(1), 34–43.

van den Eede, Y. (2015). Tracing the tracker: A postphenomenological inquiry into self-tracking technologies. In R. Rosenberger & P.-P. Verbeek (Eds.), *Postphenomenological Investigations: Essays on Human-Technology Relations* (pp. 143–158). Lanham, MD: Lexington Books.

Verbeek, P.-P. (2005). *What Things Do: Philosophical Reflections on Technology, Agency, and Design*. University Park, PA: The Pennsylvania State University Press.

———. (2008). Obstetric ultrasound and the technological mediation of morality: A postphenomenological analysis. *Human Studies*, *31*(1), 11–26.

———. (2011). *Moralizing Technology: Understanding and Designing the Morality of Things*. Chicago, IL: University of Chicago Press.

Vinck, D. (Ed.). (2003). *Everyday Engineering: An Ethnography of Design and Innovation*. Cambridge, MA: The MIT Press.

Wellner, G. (2015). *A Postphenomenological Inquiry of Cell Phones: Genealogies, Meanings, and Becomings*. Lanham, MD: Lexington Books.

Welton, D. (2006). Body and machines. In E. Selinger (Ed.), *Postphenomenology: A Critical Companion to Ihde* (pp. 197–209). Albany, NY: State University of New York Press.

Chapter 6

From Camera Obscura to fMRI

How Brain Imaging Technologies Mediate Free Will

Ciano Aydin

Brain imaging technologies, such as positron emission tomography (PET) and functional magnetic resonance imaging (fMRI), are increasingly used not only to diagnose diseases and lesions but also to correlate brain activation with psychological states and traits. Many philosophers have become interested in brain imaging technologies because the images that they display seem to have bearing on philosophical issues such as consciousness, free will, weakness of will, responsibility, moral deliberation, rational choice, agency, and identity.

In this chapter I will critically discuss the role of brain imaging technologies in free will debates. I will also briefly review the arguments of advocates and opponents of free will, focusing on how brain imaging technologies are used to support their positions. Then I will show how data and images provided by brain scanners are used to draw causal relations between brain processes and cognitive functions and discuss methodological objections against this conception. Although these methodological reflections problematize, as we will see, the view of brain images as valid and reliable representations, they neither sufficiently challenge nor explicate how brain imaging technologies instigate a framework that allows the brain to be approached as an agent that produces cognition and regulates behavior. Instead of taking brain imaging technologies as "material objects" that can neutrally represent certain faculties of the "subject," a postphenomenological methodological approach attempts to reveal the particular way these technologies frame the relation between brain and self.

Drawing upon technical mediation theory, I will investigate how brain imaging technologies mediate the conception of the brain as the locus of the self, which is then granted or denied free will. This particular notion of self and freedom has, as I will demonstrate, a history that can be traced back to Descartes: another, older technology, namely the *camera obscura*, inspired Descartes's view of the autonomous self, which is sustained and transformed by brain imaging technologies that scientists and philosophers use today in order to argue in favor of or against the existence of free will. Of course, I do not claim that the camera obscura can exclusively explain, or that it is the only source of, Descartes's views of self and freedom. However, this reconstruction will illustrate how technologies, which often remain virtually unrecognized, co-constitute seminal ideas in the history of philosophy.

The aim of this genealogical account, which can also be seen as a postphenomenological methodology, is not a plea to purify our self-identifications from technological influences but rather to show the paradoxical account of the Cartesian framework that is sustained in many free will debates: conceiving brain images as a "direct representation" of a brain that is unaffected by external influences and, hence, is seen as the locus of our selves, excludes per definition the conception that our self-identifications are technically mediated. By showing that the notion of a pure, unaffected self was co-constructed by the metaphor of a particular technology, I will argue for an alternative anthropology that properly recognizes the mediating role of technologies and propose a different interpretation of the images displayed by brain imaging technologies. This alternative anthropology will also reframe free will debates: brain scanners in themselves are not sufficient to display a brain that can be seen as the locus of the self, which presumably would render free will impossible, but could rather contribute to the formation of a self and the advancement of its freedom.

FREE WILL DEBATES AND BRAIN IMAGING TECHNOLOGIES

It is impossible, and also not my aim here, to provide an exhaustive discussion of the different positions and arguments in the free will debate. A brief account of the main arguments will be sufficient to set the stage for showing *that* brain imaging technologies inform (and, as we will see, also misinform) views on free will. It is almost undisputed that there are two minimal conditions that any action that is said to be free has to meet: the "absence of compulsion" and "the absence of pure chance." These criteria positively entail that my action is free if (and only if) it can be attributed to my (and not to an enforced or a random) will (Pauen, 2004, 2007). From this perspective we can come up with a working definition of free will: a person has free will if she, and not someone or something else, determines her will, actions, and decisions. Most advocates and opponents of free will in the context of neuroscience approve this definition of free will but disagree on the question of whether conscious intentions and decisions have causal efficacy and control in regard to the organism's behavior.

Advocates of free will attempt to demonstrate that (directly or indirectly) conscious intentions and decisions, or other aspects of our mind, can have causal efficacy and control, and are therefore not, or not completely, subject to unconscious, automatic operating processes in the brain. Since, from this perspective, we relate who we are to our mind, then what we think and do is, at least potentially, up to us. Free will is therefore possible (Mele, 2009; Libet, 2011; Baumeister, Masicampo, & Vohs, 2011; Slors, 2012, 2013).

Most opponents of free will do not repudiate the idea that there is conscious thought. What they deny is that my actions are caused by conscious intentions or decisions. They believe that the vast majority of the decisions and actions that we ascribe to our conscious intentions, including those conscious intentions, are in reality caused by unconscious brain processes. In this instance, the stance taken is that, since we relate who we are to our mind or certain contents of our mind, and our mind cannot control our brain, what we do is not up to us. Therefore, free will is, for the most radical free will critics, impossible, though we need to stress here that there are different forms and levels of "disproof" of free will (Libet et al., 1983; Hassin, Uleman, & Bargh, 2005; Wegner & Wheatley, 1999; Soon et al., 2008; Swaab, 2010; Lamme, 2010; Fried, Mukamel, & Kreiman, 2011; Harris, 2012).

What sometimes confuses matters is the claim of some of the same opponents of free will that "we are our brain." Because not consciousness but the brain determines our actions and decisions, the brain must be considered as our "real self" (Lamme, 2006, 2010). This seems to be inconsistent; if we are our brain and our brain determines our actions, then we determine our actions and are, therefore, free. That is, however, not how brain scientists like Lamme, Haynes, and Harris understand the relation between brain and self. They assume that the brain in one way or another produces consciousness, and that we (from a scientific point of view wrongly) identify ourselves with that consciousness. Since not consciousness but the brain determines our actions, and the brain—they assume—is not affected by consciousness, then the consciousness that we identify ourselves with is no more than a powerless epiphenomenon of the brain.

Both opponents and advocates of free will often strongly base their views on brain imaging technologies. Of course, neither here can I give a complete overview but rather will discuss a couple of experiments that illustrate how brain imaging technologies are used in this context. It is at least safe to say that Benjamin Libet, who has become famous for his finger wagging experiment (Libet et al., 1983), has inspired many opponents of free will. This experiment shows that the moment at which respondents indicate awareness of their decision to move their finger is slightly less than half a second after the so-called readiness potential is observed in an EEG. On the basis of this observation the (contentious) conclusion is drawn that before respondents are aware of their intention the brain has already made the decision to move the finger, which seems to undermine the concept of freewill.

Following Libet's observations, more attempts have been made to conduct increasingly sophisticated experiments and confirm that not our conscious selves but our brains determine our actions. John-Dylan Haynes's (in Soon et al., 2008) MRI set-up provided improved spatial resolution and could be used to survey the whole brain, whereas Libet's EEG technique could only record a limited area of brain activity. In this study, participants could freely decide if they wanted to press a button with their left or right hand, but were asked to remember exactly when they felt they had made up their mind, which was registered by a computer. Haynes and his team claimed to be able to predict with 60% accuracy which button subjects would choose six seconds before they were consciously aware of their decision.

Fried and his team (2011) studied individuals with electrodes implanted in their brains as part of a surgical procedure to treat epilepsy. This enabled them to record the activity in single neurons using microelectrode systems, which provided a more precise picture of brain activity than both EEG and fMRI. In their study, they showed that there was activity in a small group of neurons in particular brain areas before the subject made a conscious decision to press a button. With about 700 milliseconds to go, the researchers assert that they could predict which decision would follow with 80% accuracy.

Advocates of free will often draw upon the same brain imaging technologies but claim that they do not display the non-existence of free will. I will again limit myself to a couple of examples (see Klemm, 2010, for a more extensive overview). Some advocates of free will refer to another experiment of Libet (2011) in which respondents are given a second assignment: when they are aware of their intention to move their finger, they should try to go against that inclination. On the EEG one can see again that the readiness potential precedes the conscious intention by about half a second. However, just before the expected action we see that the readiness potential discontinues. Libet and some contemporary advocates of free will argue that it might be true that we mistakenly believe that our actions are caused by conscious intentions, but that does not exclude the ability to intervene in an impulse, and consciously veto and stop the action that our brain has unconsciously prepared. We do not have a "free will," but a "free won't."

Trevena and Miller (2010) challenge the assumption that evolving brain activity prior to conscious awareness of an intention to act is unambiguously associated with preparation for movement. According to these authors, there is no evidence that electrophysiological signs before "a decision to move" are stronger then electrophysiological signs before "a decision not to move," which would indicate that these signs are not specific to preparation of movement.

Koechlin and Hyafil (2007) have pointed out that the brain regions most frequently studied in the context of free will debates, namely the (pre-) supplementary motor area and the anterior cingulate motor areas of the brain, may be only involved in the later stages of motor planning. However, this does not exclude the possibility that other parts of the brain, which are not yet taken into account, are responsible for decision-making and exerting will.

According to Mele (2009), the readiness potential depicted by brain imaging technologies does not necessarily exclude the influence of conscious decisions on behavior. Forming certain habits could initially be the result of a conscious choice. Once habits are initiated, they may be executed with little or no conscious involvement. This explains why an intention to act in a certain way can arise without being actively formed from a decision process but does not exclude the possible influences of consciousness on our decisions and actions. This also does not exclude the possibility, as Libet suggested, that a habit may be vetoed consciously.

Although opponents and advocates provide very different and contesting interpretations of brain images, both groups often assume that data and images provided by brain scanners in one way or another depict causal relations between brain processes and cognitive functions. In the following section I will briefly examine, with reference to fMRI, what exactly brain images represent or visualize and discuss methodological objections against the view that brain images provide functional evidence.

BRAIN IMAGING TECHNOLOGIES AS FUNCTIONAL EVIDENCE?

Brain images are colorized pictures of alleged brain activity, which can be measured in different ways. fMRI, a dominant technology in brain mapping research, measures the so-called Blood Oxygenation Level Dependent (BOLD) response. When a brain area is more active it consumes more oxygen, which is delivered by an increase in blood flow in the activated area. Oxygenated and deoxygenated hemoglobin in the blood have different magnetic properties. An MRI scanner can read changes in magnetization and translate them into a computerized image (Matthews & Jezzard, 2004).

Neuroscientists and philosophers who draw a direct identity relation between brain states and cognition, use data and images provided by brain scanners to infer something about the role of particular brain areas in cognitive functions. Brain imaging data are also increasingly used, in particular in free will debates, to make a so-called reverse inference: brain imaging data are taken as evidence that a given brain region plays a particular causal role during the performance of a cognitive task.

There are "neuro-optimists" who go even one step further and assert that brain images will permit immediate access to our thoughts. Camerer et al. (2005) state, for example, that brain images can open up the "black box" of the mind to direct observation. Victor Lamme writes that an fMRI scan grants "a glimpse into someone's mind" and that we finally will be able to "measure what someone sees with his mind's eye" (2010, p. 264; see also Lamme, 2006). These views reminisce older speculations: Armstrong (1968/1993, p. 109) anticipated the idea that a "brain technician" with perfect understanding of brain processes would be able to correct the mental states that someone ascribes to himself.

A general objection against these kinds of assumptions is that brain images are not simple pictures of BOLD signal differences. It is not possible to directly measure quantitative signal magnitudes. General mapping from BOLD signal to significant neural activity is lacking (Nair, 2005). In addition, the BOLD differences that are associated to brain activity are small, noisy, and multifaceted. According to Klein (2010), brain images do not provide quantitative information but rather display *statistical* significant differences in BOLD signal between different task conditions: inferential statistics are used to contradict the so-called null hypothesis that indicates that an experimental condition had no real effect on the observed MR signal. Brain images do not directly display activation in brain areas but are rather maps of places where we can assume with a certain probability that the presumed correlation (speaking of "causation" is too premature) between data and a stereotyped pattern of activation is unlikely to be the result of chance fluctuations from a true zero signal (Klein, 2011).

Roskies (2007) draws a similar conclusion by arguing that brain images should not be conceived as photographs of brain activity. Brain imaging does not let us directly see properties carrying visual data of the brain. Unlike photography, brain imaging is belief-opaque: the information crucial to its correct interpretation is, Roskies argues, not contained in the image itself. The assumed functional decomposition, the tasks involved in producing an image and the chosen statistics are necessary for interpreting the image but not recoverable in information inherent in it. Inferences are, therefore, dependent upon the beliefs and experience of the researchers about how to analyze their results. What complicates matters here is the question of whether "direct representation" as such is possible. I acknowledge that certain conventions or beliefs might always be involved in recovering information. However, it still seems to be possible to make a distinction between more and less common or accepted conventions and beliefs, shared by a cultural community.

No less controversial is reverse inference: because brain regions can perform multiple functions and be involved in a variety of different processes, a causal relation between activation of a certain brain area and a particular cognitive function cannot be drawn (Klein, 2010). The many conceivable mappings between brain regions and cognitive functions allow for a combinatorial explosion of possible explanations of observed brain activity (Klein, 2011). Discovering more systematicity in the visual field might strengthen certain explanations (see also Klein, 2017). However, presently reverse inference provides virtually no evidence and is deductively invalid (Poldrack, 2006).

These objections against the validity of brain imaging indicate that brain images are not direct representations of brain activity and that they provide very weak functional evidence. However, for most critics these concerns about validity and reliability do not render brain imaging worthless. According to Roskies (2007), "Imaging data is, within its limits, a reliable indicator of brain activity" (p. 867). Caccioppo, Berntson, and Nusbaum (2008) believe that brain imaging can help generate hypotheses in an exploratory way, which can be tested in future studies. In a similar vein, Klein argues: "Neuroimages do not confirm functional hypothesis, but they do show brain areas in which the imaging data might be further used to confirm functional hypotheses" (2010, p. 275).

These methodological considerations and evaluations have two relevant implications within the scope of our investigation: on the one hand, they clearly show that brain images do not directly display neural activity but are, in fact, very much dependent on theoretical beliefs and statistical suppositions. On the other hand, these methodological assessments regarding validity and reliability neither challenge nor sufficiently explicate the particular way brain imaging technologies frame the relation between brain, cognition, and world.

By focusing only on methodological issues concerning functional evidence, the more basic framework that *allows* the brain to be approached as an agent that determines cognitive functions and utilizes the external world is neither revealed nor questioned. The objection that brain regions can perform multiple functions and be involved in a variety of different processes does not challenge the idea of the brain as an agent that causes cognitive functions. Although the discussed methodological reflections indicate that brain images might not be reliable and valid pictures of brain processes, they do reinforce the view of the brain as a *causal agent* by framing the brain as an isolated realm that—if certain methodological problems are solved—could provide functional hypotheses. This presumption has, as we will see below, far-reaching implications for how "free will" is conceived. A postphenomenological methodological approach allows revealing and challenging the particular way brain imaging technologies frame the relation between brain and self.

THE TECHNICALLY MEDIATED SELF

The idea that technologies frame our notions and experiences of both the world and ourselves forms the core of technical mediation theory (Verbeek, 2005). From this perspective technologies are not neutral tools that measure certain conditions and properties of the world and ourselves but they co-determine what they measure. Let me illustrate the general idea behind technical mediation theory with an example: "health" greatly derives its meaning from technologically generated measurements of variables like weight, heart rate, blood pressure, and diabetes. Today more and more people use apps to watch their health: running distance trackers, calorie counters, nutrition assessors, sleep cycle analyzers, and so on. These technologies are not neutral tools that merely measure "health," in fact "health" is codetermined by what these tools measure (and what is not disclosed and left out). They provide norms for what is considered as "healthy" and influence whether people consider and experience themselves as healthy.

From a technical mediation perspective brain imaging technologies are, as I will show, in a similar vein, not neutral instruments but frame the brain and its relations to cognition, self and world in a particular way. Consequently, through what they (presume to) measure (and what is not disclosed and left out) they strongly influence how "free will" is approached and understood.

I will prepare this analysis by drawing a parallel between Descartes's account of the relation between consciousness/self and freedom, which was, as I will illustrate, also influenced by a particular technology—namely the camera obscura—and assumptions regarding the relation between brain/self and freedom in free will debates. Next, I will discuss similarities but also one important difference between both technologies' framing of the notion of freedom. Finally, I will discuss why the notion of "freedom" that brain imaging technologies bring about is problematic.

The Camera Obscura, the Self, and Autonomy

How an idea, view or, theory emerges is often considered to be part of the "context of discovery" and is not seen as important or relevant for its validation. I believe that in certain cases recognizing the "context of discovery" can help to uncover relevant presuppositions of the theory in question and indirectly provide cues that could be used to challenge its "context of justification."[1]

Recognizing how the camera obscura played a role in Descartes's "context of discovery" will help to uncover an assumption that not only greatly determined his view of the self but also has been preserved in the view of the self that is often supposed in free will debates. Although most neuroscientists consider, in contrast to Descartes, the brain and not consciousness as the center of the self and the regulator of our cognitions and actions, there are certain similarities relating to notions of self and freedom that, from my perspective, are transferred.

I am not the first person to highlight the connection between technology and Descartes's view of the self. Lee W. Bailey (1989) argues that the camera obscura is one of the root metaphors that helped generate the image of the mind as an inner realm, ontologically separated from the external world, an image that is characteristic of our modern suppositions of subjectivity and autonomy. Technical mediation theorists Petran Kockelkoren (2003) and Don Ihde (2007) have explored this view along similar lines.

In the 16th century the camera obscura was widely used for perspective drawing. Descartes was very familiar with this piece of technology. In his *Optics*, he describes it and uses it as a metaphor to explain how the world enters our mind through our senses:

> The objects we look at do imprint very perfect images on the back of our eyes. Some people have very ingeniously explained this already, by comparison with the images that appear in a chamber, when having it completely closed except for a single hole, and having put in front of this hole a glass in the form of a lens, we stretch behind, at a specific distance, a white cloth on which the light that comes from the objects outside forms these images. For they say that this chamber represents the eye; this hole, the pupil; this lens, the crystalline humour, or rather, all those parts of the eye which cause some refraction; and this cloth, the interior membrane, which is composed of the extremities of the optic nerve. (Descartes, 1965, p. 91)

Descartes argues that in methodical self-contemplation, the self discovers that our access to the contents of our thoughts (introspection) radically differs from how we encounter the external world of material things (extrospection): we can have an authoritative (others cannot challenge our belief of being in a particular mental state), privileged (we know the contents of our own minds always better than we know the contents of the minds of other people), and immediate (knowledge of our mental contents is non-inferential and non-evidence based) access to the contents of our thoughts, which we lack toward the external world of material things. Extrospective beliefs about the outside world are indirect and dependent on deceptive sense experience (Macdonald, 2007).

This capacity makes the fundamental distinction that Descartes draws between our thinking (*res cogitans*) and the material world (*res extensa*) appear reasonable: since introspective beliefs about my mind are radically different from my beliefs about the outside world, my mind must be a completely different substance than the outside world, including my body. This argument has also another implication: since we have this privileged access to our mind and not to our body and the world, which makes us doubt their existence, we identify ourselves primarily with our conscious thought: I think, therefore, I am (*Cogito ergo sum;* Descartes, 1641/1985). Moreover, the mind is, says Descartes, in one way or another able to exert control over brain, body, and environment via the pineal gland, whereas the body is a machine that is subject to mechanistic laws (Lokhorst, 2008). The sense of being an autonomous agent is derived from the *cogito*, which is ontologically separated from the external world but in some way capable of controlling it.

The imperative prejudice that Descartes did not put in brackets in his methodical doubt experiment is that it is altogether *possible* to fundamentally detach oneself from the world and evaluate different ideas on the basis of clarity and distinctness. Although this assumption has been extensively challenged throughout the history of philosophy, especially from phenomenological perspectives (intentionality, being-in-the-world, etc.), not much attention has been given to the material/technological conditions that made this view appear cogent. Bailey and others show that the camera obscura has influenced this view because its central perspective makes the idea of an aloof spectator position, as opposed to a geometrical *res extensa*, a plausible starting point for reasoning. The camera obscura is, thus, not an innocent and neutral metaphor that Descartes used to explain his epistemology but mediated the idea of the mind as an internal representation of a world that can be observed (and acted on) by a little autonomous spectator, a homunculus situated somewhere outside the world.

This particular framing of the relation between mind, world, and self has also far-reaching consequences for conceptions of freedom. The metaphor of the camera obscura reflects the idea of freedom as a quality that is characteristic of an agent that is independent of the world and the laws and mechanisms that determine it, and, at the same time, has control over it on the basis of the images that are reflected on the screen of its (the agent's) internal control room. This framework, inspired by the camera obscura, does not mark a historical event that we have left behind, but one that has been transported and rehabilitated by other technologies, such as contemporary brain imaging technologies, which currently have become in the context of neuroscience the dominant mediating technologies on the basis of which the self is granted or denied free will.

From Camera Obscura to fMRI

It must be clear that many participants in current free will debates believe that not consciousness but the brain is the locus of the self and regulator of behavior. Although Descartes would never have situated cognition in or reduced it to brain processes, since brain and body are also part of the *res extensa*, the division between "internal" and "external" that breaks along the line of an independent agent is, as Rouse (1996) has pointed out, a Cartesian legacy.

What is relevant here is that brain imaging technologies mediate and transfer similar Cartesian notions of self and freedom. By focusing on the inside of the head and (interactions between) certain regions of brain activity an fMRI image depicts the brain as an isolated realm and allows approaching it as an independent inner agent (see also the thought experiment discussed in Dorris [2015, 37–38], in which this is taken a step further). This way a categorical distinction between "inside" and "outside" is preserved. The brain is attributed a privileged status because of its presumed capacity to detach itself from the outside world, process information that is collected by the senses, and determine, as a kind of *causa sui*, our decisions and actions.

This view of the self not only represents a Cartesian anthropology but also evokes particular notions of "control" and "freedom." Advocates and opponents of free will do not agree on the question of whether consciousness can be attributed causal efficacy (see also the entry "epiphenomenalism" in *Stanford Encyclopedia*). However, they often share the view that freedom can only be attributed to an agent (consciousness or brain) that is detached or has the capacity to detach itself from its environment or other external influences. They believe that freedom can only exist in a separate, autonomous realm, that is, a realm that is not fundamentally affected or determined by external factors. Only then is it possible to process external input by virtue of which behavior can be controlled and the external world can be utilized.

There is for our approach one important and relevant difference between the camera obscura and fMRI. The camera obscura frames how a subject acquires knowledge of self and world. In Descartes's epistemology, however, it remains rather unclear how this knowledge is obtained and used to form decisions and execute actions. The cogito is a "black box" that is only accessible through subjective introspection. It cannot be scientifically objectified and investigated (in the modern sense of the word). In contrast, fMRI has, according to optimistic neuroscientists and philosophers/psychologists (like Haynes and Lamme) who think that this technology can settle free will debates, the potential to eventually display the interior of the "dark room" and to disclose not only *what* really determines our mind and behavior but also *how* it is able to do it. From this perspective knowledge of the "real agent" of our actions and decisions—that is, of the brain—is no longer based on speculative, subjective assumptions but can be scientifically investigated and validated.

The assumption that fMRI can provide objective knowledge of how the brain causes decisions and actions does not weaken the camera obscura framework of freedom but rather reinforces it and makes it more credible. The idea of freedom as "not being determined by something else but capable of determining behavior and utilizing the world," which is now attributed to the brain—and instead of "free will" often is translated as "control"—appears more credible because it is investigated in a scientific setting.

The Illusion of a Pure, Unmediated Self

If technologies mediate to a great extent our conceptions and experiences of the world and ourselves, what then is wrong with the conception of self and freedom/control suggested by brain imaging technologies? It must be clear that the problem is not that this conception is technically mediated. The problem lies in (1) the *particular* self-identification and self-interpretation that brain imaging technologies encourage and (2) the insufficient acknowledgment of and accounting for the mediating role of these technologies in self-identification and self-interpretation processes. Both impediments are, as we will see, related.

I will first discuss and challenge the *particular* self-identification and self-interpretation that brain imaging technologies generate. Brain researchers and philosophers who identify "our real self" with the brain and attribute to the brain causal efficacy and control conceive the brain, relying on fMRI images, as some kind of *causa sui* that is detached from the outside world and can determine our decisions and actions. By correlating "our real self" with the brain, detaching it from the outside world and approaching it as an inner regulator of behavior, they rehabilitate a problematic "inside-outside" distinction. This distinction has been challenged from different "externalist approaches" that attempt to show that what we consider our "real, inner self" is greatly dependent on factors in the "outside" world. An important implication of these challenges is that the criteria for what it means to be an "independent, inner self" are blurred and, hence, the very distinction between "internal" and "external" becomes opaque (see Dennett, 2003; Hutchins, 2011; Kirchhoff, 2012).

It must be remarked that contemporary externalist accounts often concern the status of mental contents and their influence on our behavior, However, their arguments still affect positions that correlate the self or "real self" to the brain (see Aydin, 2015). Taking the brain, instead of consciousness, as some kind of inner *homunculus* (or a multitude of *homunculi*) that instigates and regulates cognitive processes and utilizes its external environment also requires clear criteria that ensure the brain (together with its physically realized mental states) is a kind of isolated "inside" that can be detached from the "outside" world (see also Rouse, 1996). These criteria are, as stated earlier, lacking.

In addition, various empirical studies challenge the view of the brain as an autonomous agent that is not fundamentally influenced by the outside world. These studies indicate that the brain is substantially affected and altered by external influences; sociocultural practices can reshape certain cortical areas of the brain or transform the brain's representational capacities (see Näätänen et al., 1997; Dehaene et al., 1999; Wheeler, 2004).

Gallagher and Crisafi (2009) and Aydin (2015) take this one step further and try to show that brain/mental contents and world are not only reciprocal but also interdependent: cognitive processes do not travel from one autonomous sphere (brain) to another (world) and vice versa (Clark, 2008, xxviii). Rather, cognition can be understood as a self-organizational process in which brains, bodies, and world simultaneously participate and depend on one another (I will elaborate on this alternative view in the next section).

The view of the brain as an autonomous agent underlies our second impediment: the effect of insufficiently taking into account the mediating role of fMRI technologies in self-identification and self-interpretation processes. The idea of the brain as an autonomous agent—a detached *causa sui*—is tenable insofar as the mediating role of the technology that co-instigates that idea remains in the background and is disregarded. It validates itself insofar as the technical conditions that have generated it remain invisible.

A parallel can be drawn again between the idea of an autonomous brain and Descartes's account of unmediated mental contents: Descartes's assumption that we can have unmediated, introspective access to our own ideas provides the alleged certainty of a self that is essentially unaffected by external influences, which presumably would secure the autonomous formation of judgments and execution of actions. In the same vein, fMRI images are believed to represent an isolated brain that is essentially unaffected by external influences, which would make its capacity to determine our decisions and actions presumably intelligible. Whether, to what extent, and in what sense mind/brain can influence our actions, I leave here in the middle (see for example Slors, 2013, for an interesting, nuanced account). What is relevant here is providing an alternative framework for understanding "free will."

Our reconstruction of the genealogy of the Cartesian self and the influence of the camera obscura on its emergence shows that even the notion of a pure, unaffected self was co-produced by the metaphor of a particular technology (see also Tenner, 2003). This genealogy highlights the impossibility of detaching our self-conceptions and self-identifications from influences of our surroundings. We tend to oversee these influences, which are increasingly technological, especially when they become part of commonly accepted practices. They then become, as Andy Clark (2003, 123) (uncritically) states, *transparent*. In the next section I will discuss how recognizing the mediating role of imaging technologies could help us to reinterpret the meaning of "free will."

RECOGNIZING THE MEDIATING ROLE OF IMAGING TECHNOLOGIES AND REINTERPRETING "FREE WILL"

From a technical mediation perspective, which is rooted in the postphenomenological approach, it is possible to recognize the structural influence of fMRI and other brain imaging technologies on our conceptions of the relation between brain, behavior and world, as well as the particular way these conceptions frame "free will." It also enables us to put forward a different interpretation of what brain imaging technologies display. This requires, first of all, an acknowledgment that brain images are not "immediate" representations or photographs of brain processes. Not only because of methodological difficulties concerning reliability and validity but also because fMRI images necessarily frame the brain in a particular way, making the very notion of "direct representation" untenable. The proposed postphenomenological methodological approach helps in revealing this particular framing. Secondly, brain images do not show that the brain is the "locus of the self" or our "real self" but rather mediate a particular conception of the relation between brain and self: brain images, in so far as they are taken as potential functional evidence, bring about the idea of the brain as an agent—and hence as a real or deep self—that instigates cognition and controls behavior. Third, brain images do not prove the existence or non-existence of free will but rather mediate a particular conception of freedom: by framing the brain as an isolated region, the (Cartesian) idea of freedom as an independent realm that is unaffected by external influences is sustained and rehabilitated.

If technologies mediate our conception of the self, as well as the self's capacity for free will, then it does not make any sense to try to acquire an unmediated, original conception of free will. We should rather acknowledge that our view of free will is and always has been influenced by the technologies that we employ to identify it. The definition of "free will" as "the capacity of an agent to determine itself and not be determined by someone or something else" disregards the fact that the medium (in our case fMRI) that is expected to establish whether "free will" exists (i.e., whether we are able to determine ourselves) co-determines how we conceive "free will." However, brain images cannot provide us with direct access to a "within"—the brain—that is not affected by a "without" because our very notion of that "within" is codetermined by external influences. The brain as an independent realm that is unaffected by external influences can only be validated in the context of neuroscience by making the external, technical conditions that have generated it invisible.

Recognizing the mediating role of brain imaging technologies in self-identification and self-interpretation processes has a bearing on the positions of advocates and opponents of free will. The idea that fMRI images could render free will an illusion is brought about by framing the brain as an isolated agent that is attributed causal efficacy, which is then considered as the locus of the self. Since fMRI images cannot provide us direct access to a "within"—in this case to a brain as an isolated and autonomous agent that determines both mental contents and bodily movements—depicting the brain as an irreducible, original, and unaffected locus of the self and the sole determinant of our behavior is impossible. The same argument applies to advocates of free will who attribute causal efficacy to consciousness and take consciousness as the locus of our self and the sole determinant of our behavior. Although I do not want to deny their influence on our behavior, neither "brain" nor "mind" can be seen as an inner agent of the self that necessitates our behavior.

Does this mean that the information that fMRI images provide is completely irrelevant? Not at all! Although fMRI images do not display an isolated realm that we can consider as the locus of our self, they can display certain correlations between brain activity and behavior. Because we can critically relate to this information provided by fMRI images, we are able to ask ourselves whether we want to identify ourselves with these habits that seem to be, at least partly, instigated by certain brain patterns or mechanisms. From this perspective, fMRI images are not used to represent the "real" determinants of a person's behavior and show that she cannot control her actions and is a powerless spectator but rather practically confront the concerned person with her actions and enable her to change them in such a way that they better correspond with the way she would like to act. From this perspective the self is not an isolated, *a priori* existing entity that is "immediately" determined by something "within"—consciousness or brain—but rather continuously discovers and forms itself by virtue of its identifications and non-identifications with its determinants. I cannot provide a full argument here but have elaborated elsewhere in more detail this interpretation of "free will," not in terms of "self-determination" but rather of "self-formation."

An example that may illustrate very roughly how brain imaging technologies could be reinterpreted is "neurofeedback" (Sulzer et al., 2013). People with, for example ADHD or autism receive, via real-time displays of EEG or fMRI, feedback on the way they respond to certain situations. That self-reflection enables them to actively influence their brain activity in order to attend in a more focused manner or respond more sociably using video or audio. Of course time will tell whether this technology is really effective. It shows, however, how brain imaging technologies could contribute to the formation of a self that better satisfies our ambitions and life projects. Since technologies mediate our self-identification, they can be integrated in our practices in such a way that they can help to bridge possible tensions between how we act and how we would like to act.

This approach puts free will in a different light. If there is no independent self that coincides with something "within" (consciousness or brain), then there also cannot be an *a priori* existing "free" or "unfree" self that is completely dependent on that "within." Acting freely in the sense of acting in such a way that *I* act, may well not have the implication that my action is the immediate result of conscious intentions, it, does, however from the proposed perspective mean that critical self-identification allows me to work on my undesirable habits and to gradually become the self that I want to be (which is always a "self in the making"), which might be a better indication of freedom.

NOTE

1. Since I reflect on the "context of discovery" only as a means to point at "blind spots," I disregard Kuhnian challenges to the distinction between "context of discovery" and "context of justification."

REFERENCES

Armstrong, D. M. (1993). *A Materialist Theory of the Mind.* London: Routledge. (Originally published 1968)

Aydin, C. (2015). The artifactual mind: Overcoming the "inside-outside" dualism in the extended mind thesis and recognizing the technological dimension of cognition. *Phenomenology and the Cognitive Sciences, 14*, 73–94.

Bailey, L. W. (1989). Skull's darkroom: The camera obscura and subjectivity. In P. T. Durbin (Ed.), *Philosophy of Technology: Practical, Historical and Other Dimension* (pp. 63–79). Dordrecht: Kluwer Academic Publishers.

Baumeister, R. F., Masicampo, E. J., & Vohs, K. D. (2011). Do conscious thoughts cause behavior? *Annual Review of Psychology, 62*(1), 331–361.

Caccioppo, J. T., Berntson, G. G., & Nusbaum, H. C. (2008). Neuroimaging as a new tool in the toolbox of psychological science. *Current Directions in Psychological Science, 17*(2), 62–67.

Camerer, C., Loewenstein, G., & Prelec, C. (2005). Neuroeconomics: How neuroscience can inform economics. *Journal of Economic Literature, 43*(1), 9–64.

Clark, A. (2003). *Natural-born Cyborgs: Mind, Technologies, and the Future of Human Intelligence*. Oxford: Oxford University Press.

———. (2008). *Supersizing the Mind: Embodiment, Action, and Cognitive Extension*. Oxford: Oxford University Press.

Dehaene, S., Spelke, E., Pinel, P., Stanescu, R., & Tsivkin, S. (1999). Sources of mathematical thinking: Behavioral and brain imaging evidence. *Science, 284,* 970–974.

Dennett, D. C. (2003). Who's on first?: Heterophenomenology explained. *The Journal of Consciousness Studies, 10*, 19–30.

Descartes, R. (1965). *Discourse on Method, Optics, Geometry, and Meteorology*, P. J. Olscamp (Trans.). New York, NY/Kansas City, MO: The Bobbs-Merrill Company, Inc.

———. (1985). Meditations on first philosophy. In J. Cottingham, R. Stoothoff, & D. Murdoch (Trans.), *The Philosophical Writings of Descartes* (pp. 1–62). Cambridge: Cambridge University Press. (Originally published 1641)

Dorris, J. M. (2015). *Talking to Our Selves: Reflection, Ignorance, and Agency*. Oxford: Oxford University Press.

Fried, I., Mukamel R., & Kreiman, G. (2011). Internally generated preactivation of single neurons in human medial frontal cortex predicts volition. *Neuron, 69*, 548–562.

Gallagher, S., & Crisafi, A. (2009). Mental institutions. *Topoi, 28,* 45–51.

Harris, S. (2012). *Free Will*. New York, NY: Free Press.

Hassin, R. R., Uleman J. S., & Bargh, J. A. (Eds.). (2005). *The New Unconscious*. New York: Oxford University Press.

Hutchins, E. (2011). Enculturating the supersized mind. *Philosophical Studies, 152*, 437–446.

Ihde, D. (2007). Art precedes science. In P. Kockelkoren (Ed.), *Mediated Vision* (pp. 25–37). Arnhem: ArtEZ/Veenman Publishing.

Kirchhoff, M. D. (2012). Extended cognition and fixed properties: Steps to a third-wave version of extended cognition. *Phenomenology and the Cognitive Sciences, 11,* 287–308.

Klein, C. (2010). Images are not the evidence in neuroimaging. *The British Journal for the Philosophy of Science, 61*(2), 265–278.

———. (2011). The dual track theory of moral decision-making: A critique of the neuroimaging evidence. *Neuroethics, 4*, 143–162.

———. (2017). Brain regions as difference-makers. *Philosophical Psychology, 30*(1/2), 1–20.

Klemm, R. (2010). Free will debates: Simple experiments are not so simple. *Advances in Cognitive Psychology, 6,* 47–65.

Kockelkoren P. (2003). *Technology: Art, Fairground and Theatre*. Rotterdam: NAi Publishers.

Koechlin, E., & Hyafil, A. (2007). Anterior prefrontal function and the limits of human decision-making. *Science, 318*(5850), 594–598.

Lamme, V. (2006). Towards a true neural stance on consciousness. *Trends in Cognitive Sciences, 10*(11), 494–501.

———. (2010). *De Vrije wil bestaat niet. Over wie er echt de baas is in het brein.* Amsterdam: Bert Bakker.

Libet, B. (2011). Do we have free will? In W. Sinott-Armstrong & L. Nadel (Eds.), *Conscious Will and Responsibility* (pp. 1–10). New York, NY: Oxford University Press.

Libet, B., Gleason, C. A., Wright E. W., & Pearl, D. K. (1983). Time of conscious intention to act in relation to onset of cerebral activity (readiness-potential). The unconscious initiation of a freely voluntary act. *Brain. A Journal of Neurology, 106*(3), 623–642.

Lokhorst, G.-J. (2008). Descartes and the pineal gland. In E. N. Zalta (Ed.), *The Stanford Encyclopedia of Philosophy*. http://plato.stanford.edu/archives/sum2011/entries/pineal-gland/

Macdonald, C. (2007). Introspection and authoritative self-knowledge. *Erkenntnis, 67*(2), 355–372.

Matthews, P. M., & Jezzard, P. (2004). Functional magnetic resonance imaging. *Journal of Neurology, Neurosurgery and Psychiatry, 75*, 6–12.

Mele, A. R. (2009). Free will: Theories, analysis, and data. In S. Pockett, W. P. Banks, & S. Gallagher (Eds.), *Does Consciousness Cause Behavior?* (pp. 187–205). Cambridge, MA: MIT Press.

Näätänen, R., Lehtokoski, A., Lennes, M., Cheour, M., Huotilainen, M., Iivonen, A., Vainio M., Alku P., Ilmoniemi R. J., Luuk A., Allik J., Sinkkonen J., & Alho K. (1997). Language-specific phoneme representations revealed by electric and magnetic brain responses. *Nature, 385*(30), 432–434.

Nair, D. G. (2005). About being BOLD. *Brain Research Reviews, 50,* 229–243.

Pauen, M. (2004). Does free will arise freely? *Scientific American Mind, 14*(1), 40–47.

———. (2007). Self-determination: Free will, responsibility, and determinism. *Synthesis Philosophica, 44*(2), 455–475.

Poldrack, R. A. (2006). Can cognitive processes be inferred from neuroimaging data? *Trends in Cognitive Sciences, 10,* 59–63.

Roskies, A. L. (2007). Are neuroimages like photographs of the brain? *Philosophy of Science, 74*(5), 860–872.

Rouse, J. (1996). *Engaging Science: How to Understand its Practices Philosophically*. Ithaca, NY: Cornell University Press.

Slors, M. (2012). *Dat had je gedacht! Brein, bewustzijn en vrije wil in filosofisch perspectief.* Amsterdam: Boom.

———. (2013). Conscious intending as self-programming. *Philosophical Psychology, 28*(1), 94–113.

Soon, C. S., Brass, M., Heinze, H.-J., & Haynes, J.-D. (2008). Unconscious determinants of free decisions in the human brain. *Nature Neuroscience, 11,* 543–545.

Stanford Encyclopedia of Philosophy, entry "epiphenomenalism." https://plato.stanford.edu/entries/epiphenomenalism/

Sulzer, J., Sitaram, R., Blefari, M. L., Kollias, S., Birbaumer, N., Stephan, K. E., Luft A., & Gassert, R. (2013). Neurofeedback-mediated self-regulation of the dopaminergic midbrain, 75C. *Neuroimage, 83*, 817–825.

Swaab, D. (2010). *Wij zijn ons brein. Van baarmoeder tot Alzheimer.* Amsterdam: Uitgeverij Contact.

Tenner, E. (2003). *Our Own Devices: The Past and Future of Body Technology*. New York, NY: Knopf.

Trevena J., & Miller J. (2010). Brain preparation before a voluntary action: Evidence against unconscious movement initiation. *Consciousness and Cognition, 19,* 447–456.

Verbeek, P.-P. (2005). *What Things Do: Philosophical Reflections on Technology, Agency, and Design*. University Park, PA: Pennsylvania State University Press.

Wegner, D. M., & Wheatley, T. (1999). Apparent mental causation: Sources of the experience of will. *American Psychologist, 54*(7), 480–492.

Wheeler, M. (2004). Is language the ultimate artefacts? *Language Sciences, 26,* 693–715.

Part III

ROBOTIC TECHNOLOGIES

Chapter 7

Paleoanthropology and Social Robotics

Old and New Ways in Mediating Alterity Relations

Michael Funk

INTRODUCTION

This chapter investigates the relations between social interactions and technical practice. Therefore, two examples will serve as empirical basis for the philosophical approach. On the one hand, insights of paleoanthropological research are emphasized in order to understand material cultures and the usage of handcraft tools (e.g., hand axes). Especially alterity relations—the social perspectives and interactions—play a major role in the complex processes of phylogenetic developments of mankind. There are the *old ways*, which tell something about human-world relations within histories of evolution. On the other hand, a second empirical spotlight focuses on social robotics—the *new ways* of social practice. With respect to both sides of the coin—phylogenetic development and current human-robot interactions—postphenomenological frameworks of *alterity relations* are emphasized on an epistemological level. With this I want to work out the different layers and forms of so-called quasi-others and second-person-perspectives. Therefore I raise a question, which has not been profoundly discussed within postphenomenological research yet: How are alterity relations possible and how might robots mediate social alterity?

From an anthropological point of view, technological processes mediate human perception per se, as technical practice is multi-perspective and culturally embedded. Cognitive evolution or the development of modern human anatomy (encephalization/brain development, erect posture of man, free hands for tool usage, etc.) have been linked to societal and sensory multistabilities (non-stable social interactions). Erect posture enables free hands for usage of tools or carrying of babies. But also the visual level was enhanced as the position of the head was higher, and better geographical orientation while constantly walking on two feet became possible. More complex forms of gestures and body language could be realized and served as booster for social capacities as well. Today human social capacities are—if not replaced—at least also challenged by new technological interfaces such as emotional and social interacting robots. Reflecting both the *old and new ways* of socio-technical mediation bears fruitful insights for conceptual progressions of postphenomenological frameworks. Therefore a hermeneutical epistemology will be elaborated in order to illustrate processes of knowing and understanding. Knowledge belongs to human persons, whereas robots remain certain forms of technical functionality. Introducing a systematic heuristics of different forms of knowing and its multi-perspective mediations, I develop a visualization that can serve as pragmatic template for interpreting concrete socio-technical actions.

Conceptual roots of my investigation can be found in the interpretations of technical practice, technical mediation and techno-material cultures as they have been elaborated by Don Ihde (1979), Albert Borgman (1984), Carl Mitcham (1994), Bernhard Irrgang (2001), Peter-Paul Verbeek (2005), and Mark Coeckelbergh (2014). Following Hans Achterhuis and the empirical turn in the philosophy of technology, I include the two mentioned empirical cases into my analyses (Achterhuis, 2001). An initial starting point of my investigation is the approach of postphenomenology following Don Ihde. In terms of embodied/bodily, hermeneutic, background and alterity relations, the concept of technical mediation has been introduced in his book *Technology and the Lifeworld* (Ihde, 1990). Ihde's argument is that technologies transform perceptions and experiences, and thereby play an important role in sensory human-world relations. *Instrumental Realism* describes a somehow materialistic approach, in which scientific reality is ensured (in contrast to idealistic platonic or positivist philosophy) and investigated within technical practice (Ihde, 1991). There is an innate close link between technological development and scientific research. For Ihde, Galileo with his telescope is a classic example of technical mediation and technoscience. Experiments and laboratory research, and usage of computers and computer models are other forms of technical mediations and illustrate the high impact of material embodiment on scientific research practices and methodologies. With *Postphenomenology* and *Expanding Hermeneutics*, the author further developed his approach by conceptually embedding the instrumental practice of technosciences into the philosophical approaches of empiric-pragmatic phenomenology and non-linguistic material hermeneutics (Ihde, 1993, 1998).

With my (2015) paper entitled "Post-Telescope-Postphenomenology? . . . and a Little Locomotive History," I intended to formulate some impulses for further conceptual developments. By focusing on the Holy Grail and locus classicus of postphenomenology—Galileo and his telescope within astrophysical observations—I suggested to develop a less physics- but more biology-oriented approach. Therefore I introduced the metaphorical term "post-telescope-postphenomenology" (Funk, 2015d). In this chapter, it is my aim to further develop this idea. I want to understand how mediating technologies and especially alterity relations shape human knowledge in a more detailed way. That will be my contribution to the discussion of *postphenomenological methodologies*.

First, a material hermeneutic approach of paleoanthropology will be introduced, with a special focus on social behavior, alterity, and second-person-perspectives. This might be the biology-oriented side of the coin. Second, human-robot interactions, especially in social robotics, are investigated. With this physics-oriented side of the coin, again, genuine aspects of alterity are associated. At this point I conclude by outlining the heuristical scheme of several forms of knowing and perspectives. It is my aim to structuralize different ways of technological mediations by systematically enfolding a visual template of embodied (*leibliche*) human-world relations. With this approach I do not want to justify the one and only ontological essence of mediating technologies. Moreover, I am following an empiric-pragmatic (postphenomenological) approach in combination with methodological constructivist and culturalist elements (*Methodischer Konstruktivisimus*, *Methodischer Kulturalismus*).[1] So, the heuristic scheme is not a platonic idealist theory of ontological truth. Instead, it is treated as a practical tool for interpreting contingent technological actions in concrete cultural situations. With this tool it is possible to better understand not only human-robot interactions.

OLD QUESTIONS AND NEW TECHNOLOGIES: MATERIAL HERMENEUTICS OF PALEOANTHROPOLOGICAL RESEARCH

Paleoanthropology is an area of research that employs a transdisciplinary methodology (Funk, 2015c; Mittelstraß, 2007). The basic question is a philosophical one: What is the human being? While following a biological approach in the trajectory of evolutionary theory (mutation, selection, and isolation of populations) it also includes elements of a wide range of other disciplines such as primatology or geological sciences and many more.[2]

Investigating the phylogeny (tribal history, *Stammesgeschichte*) of mankind includes both natural-scientific and cultural-scientific elements. By reconstructing the phylogenetic development during the last seven million years, paleoanthropologists need to address questions of natural and cultural evolution at the same time. Biological genesis of mankind is treated as a genuine amalgamation of nature and culture, between organic-genetic and material-social genesis[3]—with genuine methodological challenges. Linked with the natural-scientific and cultural-scientific approach is an engineering approach of molecular biology and synthetic biology.[4] Genetic technologies and genome sequencing up to bioengineering are not only applied in natural sciences. The intimate relations between knowing and acting (also within technical, social, and historical laboratory situations)[5] causes not only an introduction of biology to engineering, but also a high-tech embedding of natural sciences as such. With technosciences like these, the borderlines between constructivist engineering and natural-scientific investigations become blurred. The Human Genome Project (Junker, 2008, p. 8) or the sequencing of the Neanderthal genome[6] are milestones in the technoscientific development of paleoanthropology, and are examples for a development where engineering technics meet biology.

Laboratory research and computer models, as well as fieldwork, play a major role (Henke, 2007a, pp. 5, 8). A crucial impact of handcraft knowledge can be revealed within the usage of instruments in laboratories as well as in outdoor environments. At this point Don Ihde's and Peter-Paul Verbeek's concept of material hermeneutics (Ihde, 1998; Verbeek, 2005) is closely linked to paleoanthropological research practice. Paleoanthropologists interpret fossils and find their different contexts. Whereas historical investigations are characterized by interpreting certificates, dossiers, or documents, for prehistoric times nearly no text-like pieces can be found—written or artistic symbols rarely survived. Close to archaeology, the primary task of paleoanthropology is material interpretation. Whereas archaeologists are able to draw on the theoretical knowledge of inscriptions or cuneiform scripts, paleoanthropologists relate their material interpretations to the propositional context of evolutionary theory, geological, medical (anatomy, etc.), or primatological knowledge. Investigations about social behavior or societal-cultural developments of mankind in prehistoric and pre-archaeological times are even more difficult, as such human practices are not often fossilizing. Very few so-called *Verhaltensfossilien* (performance fossils, behavioral fossils), that tell something about practice, were excavated. For instance, footprints help us to understand the bipedal movement of early hominid ancestors (Henke, 2007a, p. 11; Schrenk, 2008, p. 8; Schrenk, 2009, p. 197; Schrenk & Müller, 2010,

p. 83). Also, scenarios of the usage of early handcraft tools are interpreted. Experimental archaeology plays a crucial role when possible usage of hand axes or other tools is emphasized. Therefore, recreated fossil tools are applied by today's skilled and experienced handcraft masters in the most realistic environments. On the basis of their practical evaluation, hypothetical conclusions about prehistoric hominid habits might be drawn (Mathieu, 2002).

Whereas paleoanthropology is a transdisciplinary biology-oriented cultural technoscience, it also includes some methodological and hermeneutic issues of classical *Geisteswissenschaften.* Close to historians, paleoanthropologists tell (hypothesis- and scenario-related) stories. A genuine characteristic is the development and evaluation of genealogical trees. Those graphic phylogenetic trees or even bushes (*Stammbaum, Stammbusch*) are visual interpretations of potential familial developments from early Australopithecinae to modern *Homo sapiens sapiens* (Henke, 2007a, p. 6f.; Schrenk, 2008, p. 30, 56; Schrenk, 2009, p. 299). Thereby, paleoanthropologists tell stories (Blumenberg, 2006; Funk, 2015e; Merker, 2013) that serve as hypothetical interpretations linked to material fossils, data of genome analyses, and other anatomical or geological facts. Within those stories, material fossils are linguistically interpreted by including self-images in form of an as-if story. Genesis of mankind is treated as if a certain phylogenetic interpretation would be true. Somehow, close to physicists, paleoanthropologists are observers relating to a specific perspective. But in contrast to physicists, paleoanthropologists observe their own human organic and cultural genesis. Interpreting hominid fossils means talking about the concrete, and interpreting scientists' hominid past at the same time. In this context, as well as in many others, language is not neutrally used. When a person talks about human past (as such), cultural and social self-images are—at least indirectly—implemented automatically.

In ordinary life this could cause strong ideological and religious conflicts. This happened, for example, in the 19th century when Charles Darwin presented his theory of evolution and thereby broke with the Christian doctrine of creation. Today, in the context of technoscientific research methodologies, the explication of self-images in paleoanthropological phylogenetic narratives is approved as a methodological axiom (Gutmann, Hertler, & Schrenk, 2010; Janich, 2006; Kambartel, 1989). The bodily observer is always present in the stories of human evolution. As paleoanthropological research is a form of technoscience, we can come to the conclusion, that concrete ways of bodily mediation are also directly or indirectly implemented into those stories. Without computer models or visualizations of genome sequences, there would be less meaningful histories of mankind. The human genome is not only a laboratory product or material interpretation, moreover, it is also part of a (controversial) narrative self-image. It is not only a thing, it is a history, a self-mediation and a self-alterity.

Thus, my first thesis is that technological mediation and narrative (linguistic, storytelling) mediation both play an equal role in the hermeneutic embodiment of paleoanthropological observers' perspectives. My second argument is that a genuine way of alterity relation becomes a methodological fundamental within this research practice: the technological and narrative position of an embodied observer is temporally and spatially altered. I like to call this concrete cultural and historical location "observer's self-alterity" (both temporal and spatial), which methodologically enables paleoanthropological research including technological and narrative mediations.

For example, consider the perspective of Svante Pääbo and Michael Tomasello, who are able to apply genome sequencing technologies or computer models systematically, in difference to Louis Leakey, who did classical field work and material interpretations before gene analytic technologies were invented.[7] The difference is not only the mediation of the concrete perception. Of course, laboratory technologies and genome sequencing technologies reveal totally new fossil microworlds, which could not be accessed without these methods. Here, the basic argument of biological mediation is close to physical mediation. When we come to temporal mediation, we will also realize the historic macroworlds because fossils tell something about millions of years of evolution. Again, temporal macroworlds like these might also be found, for example, in astrophysics. But a paleoanthropological specialty is the *observer's self-alterity*, because the genesis not only of inorganic or organic materials, life as such, or mammals is emphasized, but moreover the rise of hominid life and therefore the historical background of the observer itself is touched immediately. With a certain methodological distance, paleoanthropologists tell stories about themselves and our prehistoric roots. Thereby, a methodological distance to hominid pasts is important. This form of *observer's self-alterity* is one reason why there are strong controversies and ideological fights about the "true" history of mankind. Those histories are not only scientific hypotheses or arguments in scientific discourses, but also technologically mediated narrative self-images. We alter our past. And the self-alterity of an evolutionary biologist differs to the self-alterity of Christian religion. That's one reason why Darwin was not only scientifically but also socially forced to fight for his ideas.

Alterity relations are, in this sense, methodological distances. But "alter," "alterity," or "alteration" also means "variation." Alternate interpretations of human phylogeny enables differentiated interpretations of human evolution and thereby scientific hypothesis. There are many *observer's self-alterities* within paleoanthroplogical discourses. Many histories of mankind can be told. Many perspectives to hominid pasts are accessible. Or as Friedemann Schrenk puts it, there are at least as many phylogenetic trees as paleoanthropologists (Schrenk, 2008, p. 30). A crucial point is that alterity relations and several forms of technical mediations have both a strong impact on methodologies of current technosciences, as well as on the object: mankind. At this point *old ways* and *new ways* meet, as social behavior, cooperation, and empathy have become prominent research topics in paleoanthropological investigations during the last years (Stix, 2015; De Waal, 2015). Social interaction is interpreted as one fundamental factor of human evolution. Therefore second-person-perspectives and mirror neurons became prominent research topics as well.

To put it into trenchant words: alterity relations are also one of several causal factors of human evolution. This is my third hypothesis. Thereby, alterity as causal factor (in the sense of emergence) of hominid phylogeny includes at least three aspects:

1. social alterity in human-human interactions (social behavior, 2nd-PP, mirror neurons);
2. animistic alterity in human-environment interactions (treating trees or stones as entities with souls, natural spirits, or ghosts);
3. techno-material alterity that is cultural-bodily mediation (by using tools, hand axes, body decorations, etc.). The third case also includes rituals, for example, usage of drugs for shamanistic reasons (by ritually leaving their bodies shamans use imaginary/fantastical self-alterity in order to create visions and prophecies).

My claim in this section was related to paleoanthropology, the *old ways*. I argue that paleoanthropologists follow a material hermeneutical approach by interpreting fossils and their embedding (finding context). Fossils are included in phylogenetic trees and hypothetically implemented in a scenario (a story of the human past which itself needs to be proved by fossil evidence).

My first methodological hypothesis is that technical mediation is closely linked to narrative mediation. Fossils are also interpreted by using computer models and complex genome sequencing technologies. Those technologies reveal information about human evolution that are not accessible by the unweaponed eye—and therefore mediate the sensory perspective to our own hominid histories. At the same time there would be neither meaning nor sense to this technical information without narrative frameworks wherewith the hypothetical stories of human evolutionary developments are linguistically expressed.

With the second hypothesis I argue for a genuine alterity relation that fundamentally shapes paleoanthropological research methodology. The technologically and narratively mediated observer's position is temporally and spatially altered and depends on her/his concrete cultural and historical location. *Observer's self-alterity* is a methodological term, which describes these spatial and temporal alterations. Therefore a paleoanthropologist quasi-observes her/his own past in a fossil appearance, even when the fossil is temporally 200,000 years altered and spatially 10,000 kilometers altered (so maybe found in China or Africa, but investigated in a European or US-American laboratory). Putting it into other words, by telling scientific stories of human phylogeny, the storyteller needs to keep methodic distance to her/himself.

Due to current research, which is focusing on social behavior and differentiated causal factors of human evolution, I argue that alterity relations that might enable complex social interactions not only serve as methodological foundations of paleoanthropologist observers but also as one concrete causal factor of human evolution (and it is evident that technical mediation also plays a causal role here).

But how do technologies like social robots challenge social practice today (and thereby might become material factors of human evolution)?

OLD DREAMS AND TALKING TOOLS: SOCIAL ROBOTS MEDIATE THE STAGE

While in the previous section I described the *old ways* by presenting three hypotheses, it is time for the *new ways* now. Thereby, the three previous arguments will be supplemented by four theses related to social robots. Why confront paleoanthropology with social robotics? Because of the initial philosophical question: What is the human being?

Generally, manifold ways in which technics shape sensory perceptions can be differentiated. One example is astrophysics high-tech telescopes which technologically transform invisible space phenomena into visible computer screen images (Ihde, 2009). Micro- and macroworlds that are locked for the unweaponed eye become accessible by tools like microscopes or telescopes. Mediating technologies emphasize other senses as well, for example, acoustic technologies and musical instruments. Amplifiers, synthesizers, effects pedals, or computer software play major roles here (Ihde, 2007, 2015). In this section, I focus on social robots[8] in terms of technical mediation. There are at least as many ways of mediation as sensory modalities that are affected by robots: visual, acoustic, tactile (including proprioceptive self-sensation, maybe the most important one?), but also smell or taste. Imagine, for example, a toy robot in a nursery. The tool is bought for pedagogical reasons (and imagine a perfect world without any privacy or data security issues), like any other material object, the toy robot becomes embodied by the child in ways of whole body perception: seeing its colors and motions, listening to its sounds, grasping its surface, but also smelling and biting and tasting its appearance.

One of the main differences between industrial robots and social robots is the multistable potentiality of direct interaction in everyday life. While industrial robots could be interpreted as functional means for technical aims (e.g., the economic production of cars), social robots appear as functional means for social aims and therefore include human everyday life apart from isolated industry facilities. A constitutive form of mediation is the robot as "quasi-other" (Ihde, 1990, pp. 97–108; Coeckelbergh, 2011b)—pitch point of the *old ways* and *new ways*:

> The robot becomes a "you"—not as a stand-in for someone else (which we might call a "delegated second person") but a "you" in its own right, an artificial second-person, which has a claim on me as a social being. . . . By talking to the robot in second-person terms, they [people] also construct it as a quasi-other. (Coeckelbergh, 2011a, p. 65)

Social interactions, whether relating to "quasi-other" or real second-person-perspectives, involve a "mirror-effect": "seeing other humans and recognizing their shapes, we refer back to ourselves with the other-as-mirror" (Ihde, 2010, p. 42). As a technoscientific emulation (not recreation) of second-person-experience, in this sense social robots serve as mirrors of interaction. But also the design and development of social robots is challenged by a "mirror-effect." In the previous chapter, methodological self-alterity and self-expression were introduced as aspects of paleoanthropological research. A similar methodological form of alterity can be detected in robotics research as well: understanding mankind by rebuilding humans/constructing humanoid robots. The old dream to build artificial humans in combination with the will to learn something about ourselves becomes a characteristic of the constructivist engineering approach in humanoid robotics (Coeckelbergh, 2011b, p. 200; Decker, 2010, p. 46). But even if social robots do not necessarily provide a humanoid appearance, this argument remains valid. Social robots might look like animals, toys, or robotic chimera and could touch animistic behavior. In consequence, we might treat robots as entities with souls and lived emotions (the same way we do with teddy bears). Whether a robot looks like a human, a teddy bear, or a car, social robots are characterized by manifold capacities for social interaction. Thus, a challenge for engineers and designers is the reconstruction of social practice. The related approach includes the paradigm of understanding social interaction by rebuilding social interaction.

My fourth hypothesis is that the—explicit or implicit—consideration of social relations in engineering methodology leads to a physics- and cybernetics-oriented form of technoscientific self-alterity. Whereas paleoanthropologists want to understand mankind in terms of evolutionary biology—including genetics, physiology, and so on—roboticists try to pragmatically capture human characteristics by creating humanoid functionality. In both cases, observer's self-alterity—that is, perspective distance of the concrete bodily scientist or engineer toward his own human point of view—becomes a crucial part of the research methodology.

Social interaction includes language, words, symbols, body gestures, and so on. Social robots have capacities for verbal and gestural interaction and therefore a profound impact on human communication (Nishida, 2009, pp. 107f.). With body movement or natural language we talk to robots and robots talk to us. "In current human-robot relations, we can observe a shift from talking about robots and about human-robot relations to talking to robots" (Coeckelbergh, 2011a, p. 63). How could social robots challenge alterity relations and thereby also change human communicative competences?

One possible answer is related to the approach of methodological constructivism and culturalism (*Methodischer Konstruktivismus und Kulturalismus*).[9] Peter Janich analyzes that because of its multi-perspective situatedness within intentional actions, human communicative competences are principally not substitutable by technical systems (Janich, 1999, 2006, 2010). I follow this argument and apply it to social robots. This also means not to understand social robots as "autonomous," "intelligent," or "creative" actors, but as technical tools and aspects of means-end oriented social practice. Those tools can be considered as parts of techno-material cultures[10]—even if robots are technically more complex means than hand axes or hammers. Six forms of technical tools can be summarized:

Form 1: handcraft;
Form 2: machine;
Form 3: automat;
Form 4: embedded technical autonomy;

Form 5: technical semi-autonomy;
Form 6: autonomy.

A differentiation between these six forms is enabled by the categories "energy," "movement/process," and "control/framework." Forms 1 to 3 belong to "pre-modern and modern technologies" including early handcraft instruments, wooden and stone tools like hand axes (Form 1), which are fully controlled and applied by human actors with sensorimotor skills. Hereby the energy, movement (including routine, as well as new ways of usage), and the framework (including the end and surveillance) of a technical action are fully provided by the human body. Success of a technical performance while using hand axes depends on the sensory capacities of a person using the tool. Forms 2 and 3 include, for example, weaving machines or excavators—tools where the energy afforded for technical success comes from the artifact, and aspects of routine are implemented as well. Praxis, such as problem solving and finding creative solutions for various contingent and unintended situations, as well as setting the aims of a technical procedure, and the surveillance, remain human bodily embedded.

Forms 4 and 5 are related to "hypermodern technologies," including computers and robotics. The term hypermodern (e.g., Borgmann, 1992) means that modern, industrial technologies are not redeemed by postmodern, 20th-century developments; moreover, when it comes to technologies, modern developments are boosted and enhanced in the 20th and 21st centuries. "Hypermodern" in this sense means not a historical cut but a further step on the developmental path that started in early modernity, 16th- and 17th-century technosciences. Social robots belong to Forms 4 and 5, not to Form 6, which is a postulate or a posit of something that eventually becomes real: a tool that is totally autonomous. Probably, in such a case it might be adequate to avoid the word tool (Form 6). But current robots and technical systems will not belong to this postulate in the foreseeable non-science-fiction future.

What is the difference between Forms 4 and 5, on the one hand, and Form 6 on the other hand? In contrast to Forms 1 to 3, the Forms 4 and 5 include aspects of praxis, problem solving, and settings of aims on the part of the technical system. The success of a technical practice, especially in Form 5, is more independent from the human body. Human actors remain in the position of surveillance and intervention in case of a functional defect. For this reason, Form 4 is called "embedded technical autonomy": to make sure that *some* aspects of creative problem solving in contingent situations are emulated (not a 100% copy of human creativity or autonomy!) in the technical tool. As the means-end setting (e.g., of social robots) is generated by human actors, this category of technical tools includes the word “embedded.” It is a technical form of autonomy which enables capacities of social interaction—for example, giving a spontaneous verbal reaction to an unintended sentence of a user. But caused by epistemic reasons, this is not the same as human autonomy.

Moreover, in Form 5, tools like social robots functionally capture parts of the framework (the “embedding” of Form 4). Here, the system sets some aims of its own functionality—always under surveillance of human actors: for example, when social robots include complex user profiles based on manifold sensor data. The robot starts “learning” about its environment and thereby enables capacities for functionally finding its own aims. For instance when a user of a social robot often forgets his house door key, the robot might include this issue in the user profile and independently start some games or exercises with the user in order to train his skills in memory and attentiveness. Again, surveillance and intervention in case of dysfunction are related to human bodily actors. This Form 5 is called “technical semi-autonomy” as it is not a replacement of human intelligence, creativity, or autonomy, but a technical functionality which includes the emulation of some means-end capacities. Form 6, in contrast, is the hypothetical postulate of total autonomy in a technical entity.[11] Here, all (possible) sensory layers and cognitive domains of human bodies would be represented in the “tool.” This is science fiction, but not totally inconceivable.

The table[12] in Figure 7.1 illustrates my fifth hypothesis and should summarize the abovementioned differences between the tool *hand axe* and the tool *social robot*. It would be interesting and intellectually profitable to further

<table>
<tr><td colspan="2" rowspan="2">forms of technical tools (1-6)</td><td rowspan="2">energy</td><td colspan="2">movement / process</td><td colspan="2">control / framework</td></tr>
<tr><td>poiesis (routine)</td><td>praxis (problem solving)</td><td>aim / advise / end</td><td>surveillance / intervention</td></tr>
<tr><td rowspan="3">“pre-modern” & “modern” technologies</td><td>1 handcraft</td><td colspan="5">human body / actor (e.g. hand axes)</td></tr>
<tr><td>2 machine</td><td>tool</td><td colspan="4">human body / actor</td></tr>
<tr><td>3 automat</td><td rowspan="4"></td><td>tool</td><td colspan="3">human body / actor</td></tr>
<tr><td rowspan="2">“hypermodern” technologies: computer & robotics</td><td>4 embedded technical autonomy</td><td rowspan="3"></td><td rowspan="3">tool
social robots</td><td colspan="2">human body / actor</td></tr>
<tr><td>5 technical semi-autonomy</td><td rowspan="2">tool
social robots</td><td>human body / actor</td></tr>
<tr><td>postulate</td><td>6 autonomy</td><td>“tool” (?)</td></tr>
</table>

Figure 7.1 Forms of technical tools, including hand axes and social robots (Michael Funk).

develop this approach including the several meanings of autonomy. But in the scope of this paper, it is more important to see that social robots are technical tools (Forms 4 and 5) and to further focus on the epistemic ways in which these technical systems/tools have an impact on alterity relations and mediate human communication. Therefore, social robots are interpreted as aspects of techno-material culture, where robots mediate sensory perception and bodily practice in particular *situations*.[13] With a *heuristics*,[14] the epistemological and multi-perspective range of these mediations can be brought in the form of a *lucid depiction*.[15] Philosophical depictions, like the following, can serve as hypotheses as they enable hermeneutical interpretations of concrete technical practice. So, this is my sixth hypothesis[16] (see Figure 7.2).

The basic idea of the scheme in Figure 7.2 is to combine several forms of knowing with perspectives in order to schematically illustrate processes of knowing. It serves as a toolbox and template for the interpretation of concrete situations. Social performance plays a crucial role in human knowing, therefore the heuristics does not reduce knowledge to an abstract matter of

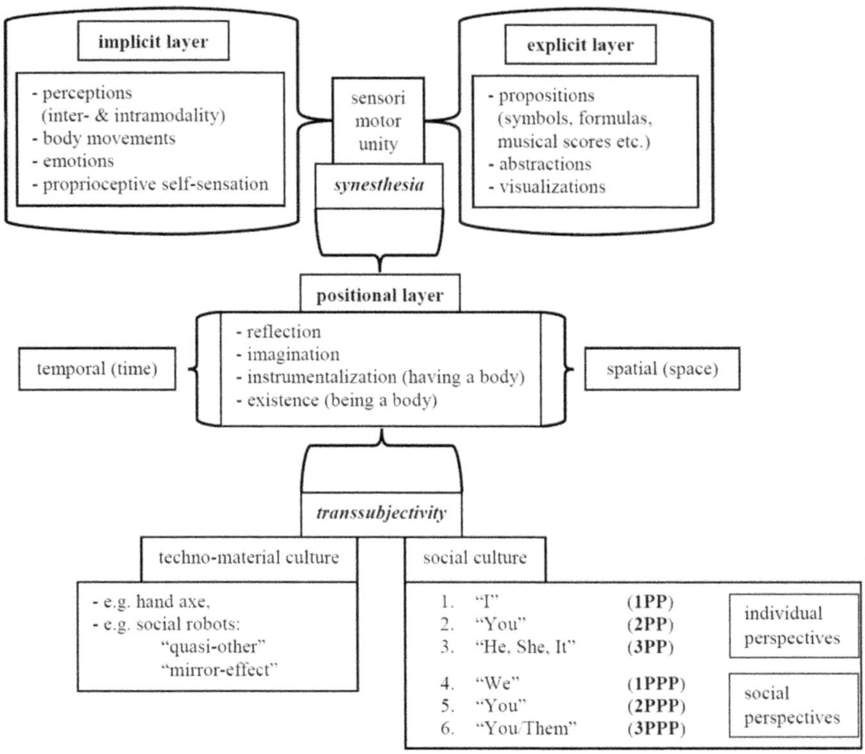

Figure 7.2 Forms of knowledge and multi-perspective bodily interactions (Michael Funk).

disembodied isolated subjects. Instead, cultural and material embodiment (*Leiblichkeit*)[17] becomes a conceptual starting point for three epistemic layers:

- Implicit layer: tacit knowledge, intuition, skills, perceptions (inter- and intra-modality), body movement, emotions, and proprioceptive self-sensation (e.g., Jean Piaget, Michael Polanyi, Hubert Dreyfus, Eugene S. Ferguson, Helmuth Plessner, but also Aristotle, and even Plato in some dialogues);
- Explicit layer: propositions, abstractions, and visualizations (like geometrical proofs) (e.g., Plato, Immanuel Kant, Gottlob Frege);
- Positional layer: multi-perspective reflection, imagination (e.g., Boethius, Augustin, Cusannus, Kant, Friedrich Kaulbach), and the double aspect of having a body (*Körper*) while existing as body (*Leib*) (Helmuth Plessner).

Technical mediation could affect each layer, or all of them, in a particular way. While a deep bass tone mediated by an amplifier affects our proprioceptive (mostly tactile) self-sensation—you can feel it in the stomach and groove with the whole body without explicitly knowing, what the muscles are doing—social robots mediate the positional layer. The conceptually most important point—also for social robotics—is the double aspect of having body and being body. This concept was investigated by Helmuth Plessner since the 1920s and became influential, especially in German *Leibphilosophie* (philosophy of human lived body): "Ein Mensch *ist* immer zugleich Leib . . . und *hat* diesen Leib als diesen Körper" (Plessner, 1941/2003a, p. 238). That means that, in an existential way, we are our bodily actions and at the same time access our body in an instrumental way. We have it like a tool—always mediated by a concrete organic and bodily position—and we exist as it (both at the same time).

"Nur in der Vermittlung durch meinen Körper, der ich selbst leibhaft bin (obwohl ich—ihn habe), ist das Ich bei den Dingen, schauend und handelnd" (Plessner, 1941/2003a, p. 247). Plessners point is that mediation starts with the human body. Even if we use tools, to some extent the double relation remains active: we are our body-tool-relation in existential ways and at the same time we keep distance to body and tool in order to instrumentalize it. Again, music serves as an example. When we play a piano, at a certain level of exercise and practice we become in unity with it. Emotions are expressed and mediated both by our body gestures and the technical instrument. But at the same time we also keep distance—so to say intentional alterity—to the piano. The distance allows us to follow certain intentions like "becoming better in blues style" or "trying Beethoven scores instead of Chopin this time." In order to realize those intentions we keep distance to our body-piano relation, exploit our body, and start learning new skills. Without this intentional alterity, we could not generate new competences. But at the same time the trial and error feedback of the environment is important. At this point, social robots enter/mediate the stage.

Social robots, as aspects of techno-material culture, provide trial and error. Capabilities for interaction with gesture-like movements or in natural languages are characteristic for social robots, in contrast to hand axes, and so on. Thereby, the concepts of *synesthesia* and *sensorimotor unity* play a crucial role. The alterity of social robots is mediated by a synesthetic unity that combines the three domains of human knowing: implicit layer, explicit layer, and positional layer. Unity is primarily ensured by sensorimotor actions, where perceptions, movements, emotions, proprioceptive self-sensations, and propositional aspects are interrelated with whole body performances. The argument of a sensorimotor lifeworld unity (in contrast to abstract idealistic unity) has been developed by Helmuth Plessner (1970/2003b), but also applied by Jean Piaget and the enactive approach of Varela, Thompson, and Rosch (1997) and other authors: "Die Lösung des Problems ergibt sich nur in Anwendung des Prinzips der sensomotorischen Einheit oder Aktionsrelativität der Sinne, wohlgemerkt unter Berücksichtigung der dem Menschen eigenen Art von Aktion," (Plessner, 1970/2003b, p. 388).[18]

The term *synesthesia* in the heuristical scheme of this chapter is used in a wide meaning. It summarizes many aspects of the linking of perceptions, symbols, movements, images, and so on. As medical phenomenon, so-called genuine synesthesia describes a very seldom habit (e.g., when a person obligatory associates a number with a certain color or the taste of coffee with the sound of a violin). But my understanding of synesthesia in this epistemological scheme appreciates a general human feature that usually remains implicit in the cognitive background. Social robots do not feature synesthesia and no sensorimotor unity as well.

My seventh hypothesis holds that tools of Forms 4 and 5 change the ways in which humans organically embed alterity into synesthetic sensorimotor unity. To put it another way, social robots, especially on the level of lingual interaction, touch alterity and thereby challenge the ways in which we socially exist as bodies and the ways in which we instrumentalize our bodies.

Time and place shape the concrete point of view. Moreover, in the heuristic scheme, the positional layer is related to techno-material culture and social culture which both belong to transsubjectivity. The concept of *transsubjectivity* (*Transsubjektivität*) is related to the approach of methodological constructivism (*Methodischer Konstruktivismus*). It has been developed by Paul Lorenzen (1985) and Jürgen Mittelstraß (1980/1984/1995/1996/2004) in order to describe the (ethically important) circumstance, that we as humans always live in shared social interactions and languages. There is no subject without reasonable community (Kambartel, 1989, p. 25; Kambartel, 1996). Insofar, transsubjectivity is the basis of all three layers: epistemic position, implicit knowing, and explicit knowing. I doubt social robots will replace transsubjectivity as such, but, as tools of Forms 4 and 5, they are aspects of techno-material cultures with strong impact on human capacities of communication. With the distinction in *techno-material culture* and *social culture,* I pay attention to the idea that the human body includes both a material (Maurice Merleau-Ponty) and a social (Michel Foucault) dimension (Ihde, 2002). Hand axes (Form 1) and social robots (Forms 4 & 5) are sensory multistable parts of human interaction. Hand axes could be used for gestural communication. Social robots talk to us immediately as "quasi-others" causing a "mirror-effect." Thus, a certain technically mediated alterity relation can be revealed with a clear impact on the positional layer of human self-experience and extrinsic-experience (including both layers of implicit and explicit knowing).

Therefore a phenomenological sketch could look like this:

organism—environment

Which means concrete:

human—robot—world

including the cases:

human—robot—technology (e.g., another robot)

human—robot—human

human—robot—lived and non-lived environment (e.g., a dog or kennel)

For relations in social practice a scheme like this might be applied:

human—social gesture[19]*—social robot*

As human bodies are shaped by a sensorimotor synesthetic unity embedded in transsubjectivity, a phenomenological structure of alterity could look like this:

transsubjectivity—synesthesia—social gesture—social robot—social gesture—synesthesia—transsubjectivity

This might be a *new way*, or more precisely a new feedback loop, of mediating techno-human cultures. The *old way,* for example, of hand axes, is somehow similar insofar as Form 1 tools also mediate experiences and can be embedded in human gestures to express a meaning. On the other hand, there are also large differences. Hand axes do not give a verbal response, even if someone is angry and throws it away. There is no "quasi-other" or "mirror-effect" with hand axes or handcraft tools.[20]

In this section, I argued that social robots, just like milk and cultures of breeding cows or goats, will become factors of material cultures and therefore will access the complex causal area of human evolution. The impact, especially on social behavior, might be bigger than with handcraft tools, like hand axes. My claim is that social robots do not replace social interaction but mediate it in *new ways* of verbal interaction. The competences we learn while interacting with social robots will influence the competences and practices by which we interact with real human beings. In order to illustrate, I summarize several phenomenological relations at the end of this section.

In my seventh hypothesis, I argued that social robots will touch the ways in which humans organically embed alterity in social life. Synesthetic sensorimotor unity, and the ways in which we socially exist as bodies, and the ways in which we instrumentalize our bodies, will be challenged at the same time. Robots become a causal (autopoiesis-like, emergence-like) factor of social behavior and human natural history (not as new or better humans, but as environmental material-cultural circumstance that provokes human creative adaptation).

Before that step, as a sixth hypothesis, I introduced an epistemic heuristically scheme in order to sketch epistemic issues of human-world relations. This scheme shall serve as a template for interpreting concrete situations of human-robot interactions and the many ways in which social robots affect complex capacities of human knowing.

In a previous fifth hypothesis, I initialized the sixth and seventh arguments by interpreting social robots not as human-like, but as tools and means for human ends. I differentiate six forms of technical tools and illustrate the conceptual differences of hand axes (Form 1, *old ways*) and social robots (Forms 4 & 5, *new ways*).

CONCLUSION: TECHNICAL MEDIATION, ALTERITY, AND KNOWING

In this contribution, subsequent to an introduction, paleoanthropology and social robotics have been discussed in two separate sections. Conceptual links have been drawn by interpreting both cases as examples for mediation in current technosciences. *Old ways* meet *new ways,* where paleoanthropologists investigate social behavior as a factor of human evolution and roboticists also emphasize social life by trying to reconstruct it. A special focus of this chapter was on alterity relations and their impact both on research methodology and everyday life. Therefore, the approach of postphenomenology has been combined with methodological and cultural constructivism. I developed seven theses:

1. Paleoanthropologists are hermeneutically embodied observers and interpret human evolution both technologically mediated and narratively mediated (linguistically, story-telling).
2. (Biology-oriented) "observer's self-alterity" becomes a methodological fundamental within paleoanthropological research practice. Biologists tell stories from technological and narrative points of view, from a concrete cultural and historical location, that is temporally and spatially alterated. They do not stand inside the last seven million years of phylogeny and cannot enter any past place at any time.
3. Alterity relations in the sense of social behavior are one of several causal factors of human evolution, including (a) social alterity in human-human interactions, (b) animistic alterity in human-environment interactions, (c) techno-material alterity that is cultural-bodily mediation (tool use).
4. Roboticists try to rebuilt capacities of social interactions and thereby follow a physics- and cybernetics-oriented form of technoscientific self-alterity as methodological fundamental.
5. A table of six forms of technical tools, where hand axes belong to Form 1 and social robots belong to Forms 4 and 5.
6. A heuristic scheme of multi-perspective forms of knowing (*Leib*) including the implicit, explicit and positional layer, synesthesia and sensorimotor unity, but also techno-material culture and social culture as issues of transsubjectivity.
7. Especially talking and gestural interacting social robots (Form 4 and 5) change the ways in which we socially exist as bodies and the ways in which we instrumentalize our bodies as "quasi-other."

More research needs to be done. Many more *old ways* meet many more *new ways* in many more ways. So, this paper is not the philosophical end to alterity relations and mediating technologies; but I hope to have provided some fruitful ideas and arguments to current discussions of *postphenomenological methodologies* and *new ways in mediating techno-human relationships*.

NOTES

1. The position of methodological constructivism (*Methodischer Konstruktivismus*) goes back to the works of Wilhelm Kamlah and Paul Lorenzen (also labeled as *Erlanger Konstruktivismus* in the earliest historical version). Inspired by Hugo Dingler, ideas of Martin Heidegger and Gottlob Frege-like philosophy of logics and mathematics have been combined in order to understand the practical lifeworld basis of scientific research, technologies, and rational justification since the 1960s. For

current discussion of roboethics and means-end schemes, see, for example, Grunwald (2012).

2. For more details and schematic overview of the integrated disciplinary approaches, see: Gruppe et al. (2012, pp. 22–23); Henke (2007a, pp. 4–8); Henke (2007b, pp. 4–6, 41–42); Henke & Rothe (1994, p. 3); Henke & Tattersall (2007, pp. vii–xi); Schrenk (2008, p. 7, 88); Schrenk (2009, pp. 197–199).

3. The evolution of mankind, oscillating between the spheres of nature and culture, has become an established research topic during the last years and is emphasized by Compagna (2015); Gerhardt (2010); Illies (2010); Irrgang (2009, pp. 47ff.); Janich (2010); Köchy (2010); Müller-Beck (2008, pp. 18–30); Schrenk (2008, pp. 77, 99, 122); Schrenk (2009, pp. 206f.); Toepfer (2013, pp. 92ff., 137ff.); Weigel (2010).

4. For latest investigations, see: Engelhard (2016); Hagen, Engelhard, & Toepfer (2016).

5. Technical, social, and historical issues of scientific research became more and more important in sociological and philosophical investigations since the 1920s with authors, like Gaston Bachelard, "phénoménotechniques," phenomenal techniques (Rheinberger, 2007, pp. 40ff.; Rheinberger, 2010; Rheinberger, 2012, p. 290); Ludwik Fleck, "Denkstile," stiles of thought (Fleck, 1935); Thomas Kuhn, paradigms and their dynamics (Kuhn, 1962); Bruno Latour and Steve Woolgar, sociology of laboratory practice (Latour & Woolgar, 1979); Karin Knorr-Cetina (1981), or Ian Hacking, turn to research practice (Hacking, 1983); and others.

6. As it has been realized only few years ago, on December 18, 2013, the Max Planck Institute for Evolutionary Biology announced: "An international research team led by Kay Prüfer and Svante Pääbo of the Max Planck Institute for Evolutionary Anthropology in Leipzig, Germany, has determined a high-quality genome sequence of a Neandertal woman" (https://www.mpg.de/7666848/neandertal_genome_project [last visited April 1, 2016]). See also, Schrenk, & Müller (2010, pp. 106–110).

7. See, for example, Pääbo's publications on the Neanderthal genome in the 1990s; Tomasello's book *Primate Cognition,* 1997; Leaky's book *Unveiling Man's Origins*, 1969.

8. Social robotics includes several kinds of applications, such as care robots, entertainment robots, education robots, sex robots, but also combat robots. For the ongoing discussion including conceptual analysis, ethical or legal issues, see Breazeal (2002); Lin, Abney, & Bekey (2014); Seibt, Hakli, & Nørskov (2014); Nørskov (2015); Seibt, Nørskov, & Andersen (2016).

9. See footnote 1.

10. Here, I am further developing my pre-study (Funk, 2014).

11. Terminologies and concepts of technical autonomy or semi-autonomy are treated in ongoing controversial discussions (e.g., Lin, Abney, & Bekey, 2014; Wallach & Allen, 2009). The concepts of "embedded technical autonomy," "technical semi-autonomy," and "autonomy" presented here are not a summary of these debates, but my interpretation of one plausible approach to differentiated forms of technical tools.

12. Presented and critically discussed at the 2nd TRANSOR ("Research Network for Transdisciplinary Studies in Social Robotics") workshop "The Significance of Simulation" (2015, June 18–19, in Kolding, Denmark).

13. Whole situations generate frameworks for the interpretation of practice. There is no action isolated from its context. But the contexts also differ. Insofar, no master-situation can be defined. This methodological impact has been philosophically investigated as *apriori* situation (*Situationsapriori*) (Rentsch, 1999, pp. 68ff.; Rentsch, 2003, pp. 75ff.).

14. Knowledge research includes non-reducible forms of knowing that epistemologically can be brought into heuristical schematic forms in order to create something like a philosophical toolbox (Abel, 2012, pp. 10–12).

15. Ludwig Wittgenstein developed in his late writings a methodology of philosophical depiction. Philosophy becomes a practice itself and investigates the—tacit, meaningful—grammar of bodily actions (Wittgenstein, 2006). Lucid depiction (*übersichtliche Darstellung*) becomes hereby one possible tool for clarifying the many ways in which we use words in a meaningful way (Gabriel, 1995).

16. The following is a synopsis and further development of my pre-studies (Funk, 2015a, 2015b), including ongoing research that I have presented and that has been critically discussed at the 1st, 2nd, and 3rd TRANSOR workshops (2015, January 28–30, in Aarhus, Denmark; 2015, June 18–19, in Kolding, Denmark; 2016, April 20–21, in Copenhagen, Denmark).

17. When it comes to "body," Martin Heidegger and Maurice Merleau-Ponty are prominently integrated, by such as the late Edmund Husserl, as well in the postphenomenological approach of Don Ihde. Also emphasized by Helmuth Plessner, the concept of body (*Leib*) became a particular research topic in German 20th-century philosophy. For a current discussion, see, for example, Irrgang & Rentsch (2016).

18. "A solution of the problem only results in an application of the principle of a sensorimotor unity or action relation of the senses—notabene, under consideration of a peculiar human kind of action," (Plessner, 1970/2003b, p. 388; translation by the author, Michael Funk).

19. For terminology of gesture, social gesture, and musical gesture, including epistemological analysis, see also the pre-study (Funk & Coeckelbergh, 2013).

20. Here, I exclude animistic positions and follow an anthropocentric approach.

REFERENCES

Abel, G. (2012). Knowledge research: Extending and revising epistemology. In G. Abel & J. Conant (Eds.), *Rethinking Epistemology* (pp. 1–52). Berlin/Boston, MA: De Gruyter.

Achterhuis, H. (2001). *American Philosophy of Technology: The Empirical Turn.* Bloomington, IN: Indiana University Press.

Blumenberg, H. (2006). *Die Beschreibung des Menschen.* Frankfurt am Main: Suhrkamp.

Borgmann, A. (1984). *Technology and the Character of Contemporary Life. A Philosophical Inquiry*. Chicago, IL: University of Chicago Press.

———. (1992). *Crossing the Postmodern Divide*. Chicago, IL: University of Chicago Press.

Breazeal, C. L. (2002). *Designing Sociable Robots*. Cambridge, MA: MIT Press.

Coeckelbergh, M. (2011a). You, Robot: On the linguistic construction of artificial others. *AI & Society, 26*, 61–69.

———. (2011b). Humans, animals, and robots: A phenomenological approach to human-robot relations. *International Journal of Social Robotics, 3*, 197–204.

———. (2014). Robotic appearances and forms of life: A phenomenological-hermeneutical approach to the relation between robotics and culture. In M. Funk, & B. Irrgang (Eds.), *Robotics in Germany and Japan: Philosophical and Technical Perspectives* (pp. 59–68). Frankfurt am Main: Peter Lang.

Compagna, D. (Ed.). (2015). Leben zwischen Natur und Kultur. Zur Neuaushandlung von Natur und Kultur in den Technik- und Lebenswissenschaften. Transcript.

Decker, M. (2010). Ein Abbild des Menschen: Humanoide Roboter. In M. Bölker, M. Gutmann, & W. Hesse (Eds.), *Information und Menschenbild* (pp. 41–62). Berlin & Heidelberg: Springer Verlag.

De Waal, F. (2015, April 21). Die Wurzeln der Kooperation. In *Spektrum der Wissenschaft*, 60–63.

Engelhard, M. (Ed.). (2016). *Synthetic Biology Analysed: Tools for Discussion and Evaluation*. Berlin & Heidelberg: Springer.

Fleck, L. (1935). *Entstehung und Entwicklung einer wissenschaftlichen Tatsache. Einführung in die Lehre vom Denkstil und Denkkollektiv*. Schwaabe.

Funk, M. (2014). Humanoid robots and human knowing: Perspectivity and hermeneutics in terms of material culture. In M. Funk & B. Irrgang (Eds.), *Robotics in Germany and Japan: Philosophical and Technical Perspectives* (pp. 69–87). Frankfurt am Main: Peter Lang.

———. (2015a). Zwischen Genetik und klassischer Musik. Zur Philosophie sinnlichen Wissens. In C. Asmuth & P. Remmers (Eds.), *Ästhetisches Wissen* (pp. 249–288). Berlin & Boston: Walter de Gruyter.

———. (2015b). Synesthesia in the philosophical works of Helmuth Plessner—And what we can learn from his approach today. In S. A. Day, M. J. de Córdoba Serrano, D. Riccó, J. López de la Torre, J. Jewanski, & P. A. Galera Andreu (Eds.), *Actas V Congreso Internacional de Sinestesia: Ciencia y Arte* (pp. 223–232). Jaén: IEG.

———. (2015c). Philosophie der Technik zwischen Paläoanthropologie und Evolutionsbiologie. Ein Beitrag zum Methodenproblem transdisziplinärer Forschung. In M. Funk (Ed.), *"Transdisziplinär," "Interkulturell." Technikphilosophie nach der akademischen Kleinstaaterei* (pp. 135–158). Würzburg: Königshausen & Neumann.

———. (2015d). Post-Telescope-Postphenomenology?. . . and a Little Locomotive History. In J. K. B. O. Friis & R. P. Crease (Eds.), *Technoscience and Postphenomenology. The Manhattan Papers* (pp. 105–114). Lanham, MD: Rowman Littlefield/ Lexington Books.

———. (2015e). Tatsachen—Modelle—Szenarien . . . Wie lässt sich das Wissen der Paläoanthropologie begründen? In J. Engelschalt & A. Maibaum (Eds.), *Auf der Suche nach den Tatsachen: Proceedings der 1. Tagung des Nachwuchsnetzwerks "INSIST," 22.-23. Oktober 2014* (pp. 80–97). Berlin: SSOAR. http://nbn-resolving.de/urn:nbn:de:0168-ssoar-454743

Funk, M., & Coeckelbergh, M. (2013). Is gesture knowledge? A philosophical approach to the epistemology of musical gestures. In H. De Preester (Ed.), *Moving Imagination: Explorations of Gesture and Inner Movement in the Arts* (pp. 113–132). Amsterdam/Philadelphia, PA: John Benjamins Publishing Company.

Gabriel, G. (1995). Logisches und analogisches Denken. Zum Verhältnis von wissenschaftlicher und ästhetischer Weltauffassung. In C. Demmerling, G. Gabriel, & T. Rentsch (Eds.), *Vernunft und Lebenspraxis. Philosophische Studien zu den Bedingungen einer rationalen Kultur* (pp. 157–174). Frankfurt am Main: Suhrkamp.

Gerhardt, V. (2010). Kulturelle Evolution. Philosophische Anmerkungen zu einem nicht erst seit Darwin aktuellem Programm. In V. Gerhardt & J. Nida-Rümelin (Eds.), *Evolution in Natur und Kultur* (pp. 185–204). Berlin & New York, NY: Springer.

Grunwald, A. (2012). Can robots plan, and what does the answer to this question mean? In M. Decker & M. Gutmann (Eds.), *Robo- and Informationethics: Some Fundamentals* (pp. 189–209). Münster: LIT Verlag.

Gruppe, G., Christiansen, K., Schröder, I., & Wittwer-Backofen, U. (2012). *Anthropologie: Ein einführendes Lehrbuch. 2. Auflage.* Berlin & Heidelberg: Springer Verlag.

Gutmann, M., Hertler, C., & Schrenk, F. (2010). Der Mensch als Gegenstand der Paläoanthropologie und das Problem der Szenarien. In V. Gerhardt & J. Nida-Rümelin (Eds.), *Evolution in Natur und Kultur* (pp. 135–161). Berlin & New York, NY: Springer.

Hacking, I. (1983). *Representing and Intervening: Introductory Topics in the Philosophy of Natural Science.* Cambridge: Cambridge University Press.

Hagen, K., Engelhard, M., & Toepfer, G. (Eds.). (2016). *Ambivalences of Creating Life. Societal and Philosophical Dimensions of Synthetic Biology.* Berlin & Heidelberg: Springer.

Henke, W. (2007a). Paläoanthropologie—Standortbestimmung einer innovativen Disziplin. *Archäologische Information, 30*(1), 1–23.

———. (2007b). Historical overview of Paleoanthropological research. In H. Winfried & I. Tattersall (Eds.), *Handbook of Paleoanthropology. Volume I. Principals, Methods, and Approaches* (pp. 1–56). Berlin, Heidelberg, & New York, NY: Springer Verlag.

Henke, W., & Tattersall, I. (2007). Preface to the Series/Preface to Volume I. In H. Winfried & I. Tattersall (Eds.), *Handbook of Paleoanthropology. Volume I. Principals, Methods, and Approaches* (pp. vii–xxiii). Berlin, Heidelberg, & New York, NY: Springer Verlag.

Henke, W., & Rothe, H. (1994). *Paläoanthropologie.* Berlin, Heidelberg, & New York, NY: Springer Verlag.

Ihde, D. (1979). *Technics and Praxis*. Dordrecht: Reidel Publishing.
———. (1990). *Technology and the Lifeworld: From Garden to Earth*. Indianapolis, IN: Indiana University Press.
———. (1991). *Instrumental Realism. The Interface Between Philosophy of Science and Philosophy of Technology*. Bloomington, IN: Indiana University Press.
———. (1993). *Postphenomenology: Essays in the Postmodern Context*. Evanston, IL: Northwestern University Press.
———. (1998). *Expanding Hermeneutics: Visualism in Science*. Evanston, IL: Northwestern University Press.
———. (2002). *Bodies in Technology*. Minneapolis, MN: University of Minnesota Press.
———. (2007). *Listening and Voice: Phenomenologies of Sound (2nd edition)*. Albany, NY: SUNY Press.
———. (2009). Imaging technologies. In J. K. B. Olsen, S. A. Pedersen, & V. F. Hendricks (Eds.), *A Companion to the Philosophy of Technology* (pp. 205–209). Malden, MA: Wiley-Blackwell.
———. (2010). Phenomenologists and robots. In *Embodied Technics* (pp. 37–53). Automatic Press/VIP.
———. (2015). *Acoustic Technics*. Lanham, MD: Rowman Littlefield/Lexington Books.
Ihde, D., & Selinger, E. (Eds.). (2003). *Chasing Technoscience: Matrix for Materiality*. Bloomington/Indianapolis, IN: Indiana University Press.
Illies, C. (2010). Biologie statt Philosophie? Evolutionäre Kulturerklärungen und ihre Grenzen. In J. Oehler (Ed.), *Der Mensch—Evolution, Natur und Kultur. Beiträge zu unserem heutigen Menschenbild* (pp. 213–231). Heidelberg: Springer Verlag.
Irrgang, B. (2001). *Technische Kultur. Instrumentelles Verstehen und technisches Handeln*. Paderborn.
———. (2009). *Der Leib des Menschen. Grundriss einer phänomenologisch-hermeneutischen Anthropologie*. Stuttgart: Steiner.
Irrgang, B., & Rentsch, T. (Eds.). (2016). *"Leib" in der neueren deutschen Philosophie*. Würzburg: Königshausen & Neumann.
Janich, P. (1999). Substitution kommunikativer Kompetenz? In M. Decker (Ed.), *Robotik. Einführung in eine interdisziplinäre Diskussion* (pp. 17–31). Bad Neuenahr-Ahrweiler: Europäische Akademie.
———. (2006). *Kultur und Methode. Philosophie in einer wissenschaftlich geprägten Welt*. Frankfurt am Main: Suhrkamp.
———. (2010). Zum Beispiel Werkzeuggebrauch. Die naturalistische Verkürzung des Tieres. In V. Gerhardt & J. Nida-Rümelin (Eds.), *Evolution in Natur und Kultur* (pp. 59–75). Berlin & New York, NY: Springer.
Junker, T. (2008). *Die Evolution des Menschen*. München: C. H. Beck.
Kambartel, F. (1989). Zur grammatischen Unmöglichkeit einer evolutionstheoretischen Erklärung der humanen Welt. In F. Kambartel (Ed.), *Philosophie der humanen Welt* (pp. 61–78). Frankfurt am Main: Abhandlungen.
———. (1996). Transsubjektiv/Transsubjektivität. In J. Mittelstraß (Ed.), *Enzyklopädie Philosophie und Wissenschaftstheorie. Band 4: Sp—Z* (p. 332). Stuttgart & Weimar: Verlag J. B. Metzler.

Knorr-Cetina, K. (1981). *The Manufacture of Knowledge: An Essay on the Constructivist and Contextual Nature of Science*. Oxford: Pergamon Press.

Köchy, K. (2010). Natur und Kultur in der Evolution. In V. Gerhardt & J. Nida-Rümelin (Eds.), *Evolution in Natur und Kultur* (pp. 39–58). Berlin & New York, NY: Springer.

Kuhn, T. (1962). *The Structure of Scientific Revolutions*. Chicago, IL: University of Chicago Press.

Latour, B., & Woolgar, S. (1979). *Laboratory Life: The Construction of Scientific Facts*. Beverly Hills, CA: Sage Publications.

Lin, P., Abney, K., & Bekey, G.A. (Eds.). (2014). *Robot Ethics: The Ethical and Social Implications of Robotics*. Cambridge, MA: The MIT Press.

Lorenzen, P. (1985). *Grundbegriffe technischer und politischer Kultur. Zwölf Beiträge*. Frankfurt: Suhrkamp.

Mathieu, J. R. (2002). *Experimental Archaeology—Replicating Past Objects, Behaviors, and Processes*. Oxford: Archaeopress.

Merker, B. (2013). Geschichte(n) der Paläoanthropologie. In C. Borck (Ed.), *Hans Blumenberg beobachtet. Wissenschaft, Technik und Philosophie* (pp. 111–125). München & Freiburg: Alber.

Mitcham, C. (1994). *Thinking Through Technology: The Path Between Engineering and Philosophy*. Chicago, IL: University of Chicago Press.

Mittelstraß, J. (Ed.). (2004). Enzyklopädie Philosophie und Wissenschaftstheorie, komplett broschiert ebd. Stuttgart: Metzler. (Original volumes published 1980, 1984, 1995, 1996)

Mittelstraß, J. (2007). Methodische Transdisziplinarität—Mit der Anmerkung eines Naturwissenschaftlers.*LIFIS ONLINE*, 05.11.07. [online: http://www.leibniz-institut.de/archiv/mittelstrass_05_11_07.pdf (letzter Zugriff am 05.01.2013) Zuerst erschienen in: Technologiefolgenabschätzung—Theorie und Praxis Nr. 2, 14. Jg., Juni 2005, S. 18–23.]

Müller-Beck, H. (2008). *Die Steinzeit. Der Weg des Menschen in die Geschichte. 4. A*. München: C. H. Beck Verlag.

Nishida, T. (2009). Towards robots with good will. In R. Capurro & M. Nagenborg (Eds.), *Ethics and Robotics* (pp. 105–116). Heidelberg: Akademische Verlagsgesellschaft.

Nørskov, M. (Ed.). (2015). *Social Robots: Boundaries, Potential, Challenges*. Farnham: Ashgate.

Piaget, J. (1980). *Abriß der genetischen Epistemologie*. Stuttgart: Klett-Cotta.

Plessner, H. (2003a). Lachen und Weinen. Eine Untersuchung der Grenzen menschlichen Verhaltens. In H. Plessner (Ed.), *Ausdruck und menschliche Natur. Gesammelte Schriften VII* (pp. 201–387). Frankfurt am Main: Suhrkamp. (Originally published 1941)

———. (2003b). Anthropologie der Sinne. In H. Plessner (Ed.), *Anthropologie der Sinne. Gesammelte Schriften III* (pp. 317–393). Frankfurt am Main: Suhrkamp. (Originally published 1970)

Rentsch, T. (1999). *Die Konstitution der Moralität. Transzendentale Anthropologie und praktische Philosophie*. Frankfurt am Main: Suhrkamp.

———. (2003). *Heidegger und Wittgenstein. Existenzial- und Sprachanalysen zu den Grundlagen philosophischer Anthropologie*. Stuttgart: Klett-Cotta.

Rheinberger, H.-J. (2007). *Historische Epistemologie*. Zur Einführung, Hamburg: Junius.

———. (2010). Gaston Bachelard: The concept of "phenomenotechnique." In *An Epistemology of the Concrete: Twentieth-century Histories of Life* (pp. 25–36). Durham, NC & London: Duke University Press.

———. (2012). Genesis of knowledge spaces and objects of knowledge. In G. Abel & J. Conant (Eds.), *Rethinking Epistemology, Vol. 1* (pp. 287–299). Berlin/Boston, MA: De Gruyter.

Schrenk, F. (2008). *Die Frühzeit des Menschen. Der Weg zum Homo sapiens. 5. A.* München: C.H. Beck Verlag.

———. (2009). Paläoanthropologie. In E. Bohlken & C. Thies (Eds.), *Handbuch Anthropologie. Der Mensch zwischen Natur, Kultur und Technik* (pp. 197–207). Stuttgart & Weimar: Metzler.

Schrenk, F., & Müller, S. (2010). *Die Neandertaler. Unter Mitarbeit von Christine Hemm. 2. Auflage*. München: C. H. Beck Verlag.

Seibt, J., Hakli, R., & Nørskov, M. (Eds.). (2014). *Sociable Robots and the Future of Social Relations: Proceedings of Robo-Philosophy 2014*. Amsterdam: IOS Press.

Seibt, J., Nørskov, M., & Andersen, S.S. (Eds.). (2016). *What Social Robots can and Should Do: Proceedings of Robo-Philosophy 2016 / TRANSOR 2016*. Amsterdam: IOS Press.

Stix, G. (2015, May). Gute Zusammenarbeit. In *Spektrum der Wissenschaft*, 52–59.

Toepfer, G. (2013). *Evolution*. Stuttgart: Reclam.

Varela, F. J., Thompson, E., & Rosch, E. (1997). *The Embodied Mind: Cognitive Science and Human Experience*. Cambridge, MA/London: MIT Press.

Verbeek, P.-P. (2005). *What Things Do: Philosophical Reflections on Technology, Agency, and Design*. University Park, PA: The Pennsylvania State University Press.

Wallach, W., & Allen, C. (2009). *Moral Machines: Teaching Robots Right from Wrong*. Oxford: Oxford University Press.

Weigel, S. (2010). An der Schwelle von Kultur und Natur. Epigenetik und Evolutionstheorie. In V. Gerhardt & J. Nida-Rümelin (Eds.), *Evolution in Natur und Kultur* (pp. 103–123). Berlin & New York, NY: Springer.

Wittgenstein, L. (2006). Philosophische Untersuchungen. In L. Wittgenstein (Ed.), *Werkausgabe Band 1* (pp. 225–577). Frankfurt am Main: Suhrkamp.

Chapter 8

Lost in Translation?

Getting to Grips with Multistable Technology in an Apparently Stable World

Lasse Blond and Kasper Schiølin

INTRODUCTION

The former creative director of the houses of Gucci and Yves Saint Laurent, Tom Ford, once remarked, "*Fashion* is what you buy, *style* is what you do about it." The same could be said about technology: "*Technology* is what you buy, *technological style* is what you do about it." In other words, technological style is how you embed technology in your daily life to fit your specific world. This saying also fits with Danish anthropologist Cathrine Hasse's call for better analytical concepts to understand how tools transform human activities and professional identities (Hasse, 2013, p. 96). In this chapter, we acknowledge the need for better analytical concepts, but we also emphasize the importance of understanding how culture transforms tools. We refer to the work of Don Ihde, who claims there is a need for a deeper insight into the ways in which technologies relate to cultural gestalts (Ihde, 1993, p. 26). However, while we recognize that technology shapes culture (a view also expressed by Verbeek, 2005), we seek to nuance Ihde's understanding of cultural responses to adopted technologies (Ihde, 1993, p. 28) by suggesting that culture also shapes technology (Rammert, 2002).

Reconsidering Ihde's cultural hermeneutics (Ihde, 1990), postphenomenology seems particularly well placed to comprehend the importance of the cultural context of technology. However, we suggest that Ihde's notion of the macro-perceptual field could be improved by the concept of "technological style," which would recognize that technology is shaped by external cultural factors, resulting in a distinctive style (Hughes, 1977). We would therefore like to re-emphasize the cultural level of human-technology relations and suggest that postphenomenology has placed too much emphasis on technology, leaving the mediated human "I" and the world in the dark. To compensate for this imbalance, we investigate how the human "I" and the world might be grasped within a postphenomenological framework. Recognizing the difficulty of this task, we introduce the notion of "human instability" and thereby underline the need to understand that the human mediated by technology is never a stable subject.

Our focus on the cultural context of technology naturally leads to an inquiry into what the world is. We suggest it is possible to handle this daunting philosophical question by paying attention to informants in the empirical field and their comprehension of their "apparently stable world." Thus technology, and especially the phenomenon "transfer of technology" (i.e., the relocation of a technological artifact), exposes cultural factors that are distinctive of the apparently stable world; it prompts postphenomenology to consider them and, in our case of a travelling technology, to ensure they are not lost in translation.

We recognize the incitement within postphenomenology to make the philosophy of technology more empirical. However, if postphenomenology is to progress beyond *program 1*—Ihde's basic scheme of different human-technology relations, which structures much postphenomenological analysis—the philosopher, social scientist, or contemporary historian needs to leave the office and search for humans in their worlds. By doing so, they can avoid the danger of their fascination with new technology coloring their comprehension of the human-technology relationship. Thus, we would like to highlight our analytical points by considering an empirical example of the human-technology relationship(s): the transfer of the social robot Silbot from Korea to Denmark and Finland.

TRANSFER OF TECHNOLOGY

Peter-Paul Verbeek commends Don Ihde for being the first to explicitly connect phenomenology with the philosophy of technology (Verbeek, 2005), but Ihde should also be lauded for his willingness to scrutinize the phenomenon of the transfer of technology (TOT)—a domain thoroughly explored by economists (e.g., Ruttan & Hayami, 1973; Rosenberg, 1976) and historians of technology (e.g., Lindqvist, 1984; Pacey, 1990).

What characterizes Ihde's writings about TOT is his understanding of the complexity of the phenomenon in *Technology and the Lifeworld* (Ihde, 1990, p. 128). As the economist Nathan Rosenberg so incisively identifies, TOT is not merely a question of relocating "a piece of hardware" (Rosenberg, 1976, p. 174). This is also acknowledged by Ihde, who calls for a more multidimensional and phenomenological model to understand TOT and to recognize the "basic cultural and existential interchange" as part of the phenomenon (Ihde, 1993, p. 34). However, Ihde also emphasizes that adaptation of the transferred technology is determined by its ability to be incorporated into an "extant praxis" in the recipient country. Ihde further elaborates on this essential point: "But even when it is adapted, the context of significations may differ quite radically relative to the sedimented type of praxis in the recipient culture. One does not need to go to exotic examples to take note of this phenomenon" (1990, p. 127).

Thus Ihde stresses the importance of recognizing how technology is culturally embedded (Ihde, 1993, p. 27), thereby highlighting that the cultural contexts of TOT should be taken into consideration for a more comprehensive understanding of the phenomenon. However, a more nuanced comprehension of the recipient culture is not the only aspect of TOT that needs to be accounted for, the "multistability" of technologies is just as important. In other words, focusing on the multiplicity of uses of a technological artifact and the fact that a technology may be understood differently in various cultural contexts and thus be embedded varyingly (Ihde, 1990, p. 144). Ihde remarks that the use of a technological artifact is "deeply unpredictable and uncontrollable" (Ihde, 1993, p. 37).

According to Ihde (1993, p. 39), it is inevitable that TOT causes cultural change; however, he does not focus on this aspect when exploring the phenomenon. Instead, he focuses primarily on the preconditions for TOT. For Ihde, the embedding of a transferred technology and its contact with "recognizable praxis, the familiar" (Ihde, 1993, p. 40) seems particularly important. For this reason, he never loses sight of the significant cultural context that the recipient country constitutes. The cultural level of human-technology relations is the focal point of his cultural hermeneutics and the macro-perceptual field "within which our bodily involvements take place" (Ihde, 1990, p. 124).

In his seminal work *What Things Do*, Peter-Paul Verbeek applies the postphenomenological perspective as a framework for understanding technological artifacts, and aims to expand postphenomenological philosophy by approaching Bruno Latour and his theory of technical mediation. Although he acknowledges that Latour's Actor-Network Theory and the phenomenological perspective seem an odd couple, this does not hamper his ambition to understand reality by extending the postphenomenological perspective with Latour's concepts (Verbeek, 2005, p. 162). Inevitably, this courting of Actor-Network Theory gives Verbeek's ontology an excessive focus on the mediating role of technologies. This gives rise to his idea to anticipate the consequences of the mediation of artifacts by designers inscribing ethics into their products (Verbeek, 2005, p. 218), thus materializing morality (Verbeek, 2006, p. 379).

Although profound, this focus on mediation of technology does not preclude Verbeek from recognizing the multistability of technological artifacts, which obviously complicates anticipating "the eventual character of mediation" (Verbeek, 2005, p. 217). Yet, multistability seems to yield to mediation or "technological shaping":

> Technology is never purely determinative, for in principle other cultural relations with a given artifact are always possible. But neither is it purely instrumental, for when an artifact receives a particular definition within a cultural context—and thus becomes stable rather than multistable—it still contributes to shaping that context. (Verbeek, 2005, p. 138)

Anticipating this shaping is being aware of the materialized morality (Verbeek, 2006). According to Verbeek, the question of how technology shapes culture constitutes the main focal point (Verbeek, 2005, p. 138). The question of how culture shapes technology is secondary. However, we argue that it is too narrow to focus on the technological transformation of culture, thus neglecting the *cultural shaping* of technology. We wish to explore how culture limits multistability, or, to use Peter-Paul Verbeek's words, how "conventions" come into place concerning given technological artifacts (Verbeek, 2005, p. 217).

This seems particularly relevant in relation to TOT. What is in fact at stake in TOT is a relocation of culture or a "cultural instrument" (Ihde, 1990). By examining the transfer of the South Korean robot Silbot to a Danish rehabilitation center, we aim to provide a more elaborate way of understanding how technological ensembles are adopted into praxes. However, before we do so, it is first necessary to explain the term "cultural shaping" and to introduce the term "technological style" into the postphenomenological vocabulary.

Although he does not use the term "cultural shaping," the distinguished historian of technology Thomas P. Hughes describes the phenomenon of cultural shaping in his magnum opus *Networks of Power*: "The local conditions external to the technology can be defined as cultural factors; the technology they shape, a cultural artifact" (Hughes, 1983, p. 405). In this sense, Hughes's argument is similar to Ihde's (1993, p. 13) description of technologies as "cultural instruments" that have to be embedded in daily life praxes. The German sociologist Werner Rammert also shares this view. Rammert believes that shaping is a two-way interaction, as opposed to a determined one-way course, and that cultural patterns shape the development of technology more than we realize (2002, p. 174f.). However, Rammert does not focus specifically on the cultural exchange caused by international TOT, whereas Hughes writes extensively about international

TOT (cf. Hughes, 1962, 1983, 1987, 1995). In many ways, Hughes's work on TOT complements Ihde's work, but the former elaborates our understanding of how technologies are adopted into or embedded in praxis by working with the concept of technological style. This focal point is essential for comprehending TOT and grasping the phenomenon as a two-way interaction.

Hughes introduced his concept of style in the late 1970s while writing about regional technological styles in connection to power systems in England, Germany, and the United States and recognizing that factors external to technology should be emphasized in order to explain the different shapes of these systems (Hughes, 1977, p. 230). The shape of a technology can be elucidated by taking non-technical factors into consideration; factors descriptive of the cultural context. In *Networks of Power*, Hughes defines technological style as follows: "Technological style can be defined as the technical characteristics that give a machine, process, device, or system a distinctive quality. Out of local conditions comes a technology with a distinctive style" (Hughes, 1983, p. 405).

For this reason, local conditions viewed as the cultural context should not be underestimated when trying to understand the development of technology (Pantzar & Shove, 2010, p. 456). Instead, they should be regarded as equally important aspects of TOT, because "adaptation is a response to different environments and adaptation to environment culminates in style" (Hughes, 1987, p. 68). This adaptation into or embedding in a local praxis can be seen as a cultural shaping of technology. Hughes's primary example is the transformer developed by Lucien Gaulard and John Dixon Gibbs, which had a "decidedly British style" (Hughes, 1995, p. 455). These specific British features were a result of requirements of the British Electric Lighting Act of 1882 that had resulted in a very complex transformer difficult to export—yet adaptable to several different voltages. To overcome this challenging import, the recipient countries had to adapt it to their local conditions and strip it of its British characteristics (Hughes, 1995, p. 453). For example, by redesigning the Gaulard Gibbs transformer in the United States, Westinghouse produced an American transformer with a distinctive American technological style, not having to pay attention to British legislation. This case is by no means exceptional. The history of technology is full of examples of technologies shaped by recipient cultures: For example, the steam train acquired a distinctive Russian style when transferred from the West by reinventing their cylinder system (Pacey, 1990, p. 152). Similarly, the French Fouga airplane acquired a distinctive style adapted to local cultural preferences in Israel by replacing various metal parts of the plane with fiberglass solutions in order to reduce weight and cost, which gave early indications of what Israeli technology would be like (Bloch, 2004, p. 26). Indeed, the historian of technology Arnold Pacey believes the way to avoid negative results of TOT is to create space for a dialogue that may lead to modifications (Pacey, 1990, p. viii), thus acknowledging the two dimensions of the shaping of technology.

Hughes and his associates in the field of the history of technology have introduced the idea that embedding technology in daily life praxis is understood more profoundly by incorporating the concept of technological style into the cultural hermeneutics. Hughes describes this as follows: "To see culture as a shaper of technology suggests a broader range of influences affecting technology *not simply the social*" (Hughes, 1995, p. 451; our emphasis). Thus, the cultural context of technology must be taken into consideration as a shaper of technology but also as a constraint to the multistability of a technological artifact transferred from one culture to another culminating in a culture-specific technological style. Besides from recognizing the social shaping, this implies paying proper attention to spatial and temporal aspects (Hughes, 1983, p. 405; Shove, Watson, & Pantzar, 2012, p. 123). This is particularly important if we require postphenomenology to explain how the "conventions" of use of a technology emerge (Verbeek, 2005, p. 217) and if we wish to grasp the notion that technologies continuously shape and are shaped by cultural contexts. The cultural level of human-technology relations should not be underestimated.

Finally, Sheila Jasanoff's extensive works in science, technology and technology studies (STS) should be mentioned as a resource to better understand the cultural diversity of the spaces in the world, in which humans and technologies reside and relate to one another. Comparing various technological innovations in different national settings by studying the different regulatory and legislative practices and, more recently, the different socio-technical imaginaries (Jasanoff & Kim, 2009, 2015) accompanying them,[1] Jasanoff has persistently advocated for the methodology of comparison in STS:

> Comparison thus solves the problem of the "view from nowhere"—that position of mythic neutrality that no analyst can achieve in practice. Instead, this method allows different, actual "somewhere" to be brought into productive contrast, revealing patterns and persistences that might otherwise remain unperceived. (Jasanoff, 2012, p. 7)

From this point of view, the "where" question becomes essential: Where is technology? That is, in which relatively, or at least apparently, stable "discursive tracks, laid down on historically contingent institutional foundations, and lubricated by repeated articulations for audiences attuned to specific modes of demonstration and argument" (Jasanoff, 2012, p. 19) is technology to be found? And how do these cultural tracks differ and influence the human-technology relations? If we do not want to be caught in the abstract view from nowhere, such questions cannot be ignored.

SILBOT—THE TRAVELLING ROBOT

As a way of illuminating this point further and remaining true to our postphenomenological methodology, let us examine an empirical and contemporary example of TOT: the transfer of the robot Silbot from Korea to Denmark and Finland.[2] By doing so, we acknowledge Verbeek's attempt to make the philosophy of technology empirical by trying to understand technology in terms of "concrete artifacts" (Verbeek, 2005, p. 6). In order to get a deeper understanding of the transfer of Silbot to Denmark and Finland, central participants of the Finnish and Danish project were interviewed and the original test facilities were visited in both Helsinki and Aarhus. Furthermore, extensive participant observation has clarified the continuous development and appropriation efforts undertaken by the Municipality of Aarhus to adopt the Silbot robot into local elderly and health care praxis.

The social robot Silbot was developed by the Korean Institute of Science and Technology (KIST) to teach English to school students in Korea. Originally called EngKey, the idea was to facilitate real-time interaction between students and the robot, which was controlled by the students' teacher from another room. Besides this, the director of the robotics department at KIST believed the egg-shaped robot might have the potential to treat or slow down the progression of dementia. The robot was then reprogrammed to facilitate cognitive exercises through interaction with elderly citizens by means of tablets.

The municipalities of Helsinki, Finland, and Aarhus, the second largest city in Denmark, displayed an interest in testing the robot for this purpose. In cooperation with KIST, test beds were built in a center for the elderly on the outskirts of the Finnish capital and at a center for rehabilitation in Aarhus. In these two locations, cognitive exercises were carried out with elderly volunteers from the fall of 2011 to February 2012.[3] Thereafter the pilot-tests were evaluated with varying results.

The Finnish project team did not find Silbot "reliable enough" and the robot was deemed "not ready for recreational services."[4] Although the Danish project team identified technical as well as cultural challenges, they were more lenient toward Silbot's limitations and, to this day, they continue to work with the robot in close cooperation with KIST.

A closer look at the Finnish and Danish evaluation reports[5] reveals other interesting aspects of the transfer of the Korean robot. Both parties stressed the need for a mediator between the students or elderly citizens and the robot;

the Finns found that the robot could "not replace human instructors" and the Danes found that Silbot required "a lot of intervention from the teacher in relation to the pedagogical content" and that the teacher could by "no means be dispersed." The Finns even bemoaned the fact that the robot did not support the work of the teacher and that the robot was too complicated for the teacher to use.

However, the main reasons why the TOT project failed for the Finnish team was that it was a technological "push project" that lacked the "flexibility that is sometimes needed in innovation projects"; also, Silbot was unable to "answer" the users' needs, because it was "predesigned" before arriving in Helsinki.[6] Therefore, the whole project lacked "flexibility," and instead of responding to the elderly citizens' challenges directly, the question became how the robot could help the professionals. In other words, the Finns were unable to adopt the robot into their praxis. During the interviews both users and the project team members deemed the robot unnecessary, underdeveloped, inflexible to even smaller adjustments and too expensive.

In Denmark, Silbot's use of the Danish language caused some raised eyebrows. The robot was prone to scolding the elderly citizens when they did not solve the cognitive exercises within the predetermined time frame. The robot's use of "inappropriate" and abusive language (for example, suggesting that the elderly participants were drunk when they failed to solve the cognitive games), as well as incomprehensible phrases, was also noticed. These translational problems were deemed counterproductive to Danish learning styles and a barrier to creating a positive learning environment for the elderly citizens, as was the very rigid test schedules designed by the Koreans (and similarly used at a center for Dementia in Seoul where Silbot was tested as well). These schedules dictated 90 minutes of uninterrupted cognitive training. The Danes quickly shortened the duration of these training sessions to make them less demanding for the elderly citizens and requested the cognitive games be reprogrammed in order to provide the elderly citizens a longer time to respond.

The Danes concluded that Silbot needed to be "more adjusted according to Danish culture" and that the "cultural dimension" should be seen as a "serious challenge" to be "addressed in the future."[7] However, the Danes continued to use the robot. They compensated for these challenges and reported its shortcomings to KIST, but, perhaps more importantly, they also outlined their expectations for the design of the new version (or update), the present Silbot 3.0. Thus the Danes directly shaped the Silbot robot. The present version still has grave technical challenges, but the language, the design, and the cognitive exercises have been adjusted and the Danes continue to exchange and report defects to be remedied by the South Koreans.

How could we understand the transfer and use of Silbot from a postphenomenological perspective? It seems evident that the transfer of Silbot to Finland and Denmark is an example of cultural interchange(s), which Don Ihde believes should be taken into account in TOT. In accordance with his perspective on transfer, the praxis in the recipient culture should not be underestimated; indeed, it seems crucial for the success of the transfer, understood as the cultural embedding of the transferred technology.

At this stage, it remains unclear how Silbot has shaped the culture at the elderly center in Helsinki and the rehabilitation center in Aarhus. In order to establish this, it would be necessary to gather more empirical findings at the sites. However, the effort to comprehend the transfer of Silbot to the Nordic countries underlines the importance of keeping cultural hermeneutics as a part of the postphenomenological perspective. As Ihde originally proposed, the macro-perceptual field seems essential to comprehend the phenomenon of TOT.

We would like to argue that it would also be worthwhile to include the concept of technological style in Ihde's cultural hermeneutics. Observing Silbot, it becomes clear that the robot has distinct Korean features; the design, the visual content of the cognitive games and the characteristic graphic that constitutes the "face" of the robot are clearly foreign. Did this technological style prevent Silbot being embedded in Finnish praxis? Certainly some of the elderly Finns found the exaggerated facial expressions of the robots naïve and childish and not to be expected from a robot with the duty to function as a teacher and instructor. Whereas the Koreans deemed evident facial expressions highly important in order to provide trustworthy communication.

Should the new version of the robot, which is based more closely on Danish system requirements, be seen as a case of stripping the robot of its Korean characteristics in the continuous effort to give Silbot a Danish technological style? These are central questions regarding the application of the concept of technological style in the understanding of the transfer of Silbot; however, it is difficult to provide definitive answers at this stage.

In the Danish case, it seems the conventions of use are still being negotiated and the efforts to embed the robot in praxis are continuing. Silbot has definitely been stripped of some of its Korean characteristics. The content of the cognitive games have been altered; and the robot's repertoire of songs and stories now includes tales of Danish Viking Kings, runic stones, and popular ballads to be memorized by the participating elderly citizens. Therefore, although its use in Denmark continues to be shaped by specific Danish demands and praxis, Silbot remains multistable. Thus the Silbot case can be seen as an exchange between two technological traditions or styles: the South Korean and the Danish. In any case, this two-way exchange must be recognized from a postphenomenological perspective as *culture shaping* and *cultural shaping*. The conditions external to a technology should also be considered. By elaborating the understanding of cultural shaping and introducing the concept of technological style, we aim to respond to Don Ihde's call for "a deeper insight into the ways in which the ensemble of technologies relates to cultural gestalts" (Ihde, 1993, p. 26).

BEYOND "PROGRAM 1"

We suggest that the problem of recognizing conditions external to technology in postphenomenology may be embedded in the very core of the theory, or at least in one of its most well-known programmatic trademarks: Don Idhe's (1990, p. 72ff.) *program 1*, which categorizes different human-technology relations. Through Verbeek's (2005, p. 122ff.) reception of program 1, it has indeed become the locus classicus of postphenomenology and the point of departure for many newcomers and students in the field.

The overall argument in the program is that humans always perceive the world through technology: "It is something humans have always—since they left the naked perceptions of the Garden—done" (Ihde, 1990, p. 72). Schematically, this idea is usually explained as:

I—Technology—World

Humans do not perceive the world directly but through mediatory layers; for example, they see the world through glasses or feel the world through clothes.

The installment of technology as a mediator between the human and the world has indeed transcended the somewhat rigid Cartesian dualism between the subject (*res cogitans*) and the object (*res extensa*). But while postphenomenology, through comprehensive accounts of technology, has done an excellent job providing flesh and blood to the neglected perceptual technological mediator, both the human "I" and the world still seem to lead a lifeless existence as bare Cartesian subjects and objects; as if the world were not really a *life*world, and the human, paraphrasing Schopenhauer, were "a winged cherub's head" without a body in pain, lust and desire, a language, a gender, or a nationality. Verbeek is also aware of this potential danger with human-technology relations: "By saying that mediation is located 'between' humans and world (as in the schema I–technology–world), Ihde seems to put subject and object over against one another, instead of starting from the idea that they mutually constitute each other" (Verbeek, 2005, p. 129). While Verbeek acknowledges that, in *Expanding Hermeneutics*, Ihde makes it clear that "subject and object are mutually interrelated," he also claims that Ihde "does not connect this thought with his earlier analysis of human-technology relations" (Verbeek, 2005, p. 129). We agree with Verbeek entirely on this issue. Like Verbeek, our aim is "to hone" the theory of human-technology relations.

A clear indication of postphenomenology's preference for the technological mediator is its selective use of other philosopher's works. For example, when Merleau-Ponty is mentioned, it is primarily to elaborate on his feather or his cane, not on his great insights into human psychology. The same is true with Heidegger; his persistent attempt to unfold the human *Dasein* by mapping its existential structure—its being toward death, its anxiety, its historicity—remains almost untouched, whereas the Heideggerian hammer is turned upside down. To say it simply: It seems as though postphenomenology has been blinded by its urgent task to investigate the technological mediator, and, in doing so, it renders the two mediated parts of the equation more or less unknown.

What is needed, we suggest, is an expansion of the postphenomenological program 1, driven by two fundamental questions: "What/who is the human?" and "What/where is the world?" Of course, one could object that such big questions cannot be answered at all. But then the same is true with the question "What is technology?" and this question has kept postphenomenology and the philosophy of technology occupied for years. Once in a while, it is necessary for an academic field to reflect and explicate the fundamental concepts it takes for granted; sometimes, even mathematics asks: What's a number? In postphenomenology, the human and the world are indeed such axiomatic concepts. Although we do not pretend to answer these questions, we hope to identify possible ways for postphenomenology to reflect on hitherto unquestioned aspects of its conceptual framework. We hope that this will be conceived as an invitation to discuss these questions further and elaborate on the meaning of the human and the world within a postphenomenological framework.

TOWARD A POSTPHENOMENOLOGICAL ANTHROPOLOGY: *HUMAN INSTABILITY*

The question "What is the human?" can be treated in many different ways. Within a philosophical scope, it is usually a matter of "philosophical anthropology." In the 1920s, Max Scheler (2009), Helmuth Plessener (1981), Arnold Gehlen (1988), and others established philosophical anthropology as an independent discipline that draws on insights from such different fields as biology, anthropology, psychology, and philosophy to approximate an understanding of what the human is. In a less strict sense, philosophical anthropology is practiced throughout the entire history of philosophy. Each time a philosopher tries to sketch out some fundamental features of being human, (s)he engages in philosophical anthropology. For instance, Aristotle's definition of the human as "animal rationabile" can be understood as philosophical anthropology.

However, as mentioned above, postphenomenology lacks an anthropology. In his inaugural talk at the University of Twente, Verbeek (2009) very cautiously anticipated philosophical anthropology as an important component in a contemporary philosophy of technology, so perhaps it is not far from postphenomenology to think along these lines. However, Verbeek is yet to connect philosophical anthropology directly to the human-technology relation to ask who the "I" is that encounters the world through technology.

The postphenomenological "I" does not have specific characteristics, such as a gender, an age, or feelings. However, in the human-technology relation, such specific characteristics can be significant. In the Silbot case, it is important to consider that the people interacting with Silbot are elderly and suffering from dementia. This could be one of the explanations why Silbot's language was deemed inappropriate in Denmark but not in South Korea, where Silbot was interacting with healthy school children. Age and health—and presumably many other human features—thus seem to shape the relation to technology. For instance, if one seeks to understand the relation between a human driver and a car, it is not enough to presuppose an anonymous and abstract "I." On the contrary, one must ask about the driver's age, gender, nationality, medical history, possible previous drug use, and other factors. Just as the object is transformed into full-fledged technology through postphenomenological analyses, the subject must also be provided with human traits to become a full-fledged human being.

To create such an anthropological awareness, we would like to contribute another concept to postphenomenology's analytical toolbox: "human instability." Just as the technology is always *multistable*, the human is always *unstable*. While Ihde (2012) uses the Necker cube and other technologies to show that technology is multistable—that is, that it has no stable essence—he (or Verbeek) never questions whether the human is (multi)stable. We are left with no information about the mood or state of the human looking at the Necker cube—for example, whether it is an angry old woman, a lustful young man or a sleepy child—or how different moods and states shape the human-technology relation. However, if the human-technology relation is understood as a relation between an "unstable human" and a "multistable technology," one could better appreciate how basic human features, such as moods, influence the use of technology. Moreover, the concept of instability would also explain the diverse ways in which the same person could relate to the same (multistable) technology; for instance, how a person's relation to the same computer can change by the minute when switching between serious work, online chess, or porn. When the "I" acquires an age, a sex, and is seen as lustful, playful, rational, irrational, sleepy, lively, sad, happy, or libidinous—in other words, as an unstable human being—justice is done not only to the human part in the human-technology-world scheme but also to the understanding of technology. Emphasizing the instability of the human being could thus be a first step in exploring the philosophical anthropology of postphenomenology.

A WORLD OF . . . ?

As described above, postphenomenology's notion of the world is not as underdeveloped as its notion of the human, and Ihde's focus on the Husserlian lifeworld and technology's dependence on context is indeed a good starting point to discuss "what the world is made of." Nevertheless, there seems to be some unknown factors in the postphenomenological concept of the world, which become explicit when the theory of human-technology relations is confronted with empirical findings. These unknown factors are also what Hughes is trying to highlight in the earlier cited quotation suggesting that there is a broader range of influences that affects technology than the social.

One of the reasons why the Danish rehabilitation center experienced problems embedding the Silbot robot in its daily life praxes was that Silbot needed to be "more adjusted according to Danish culture" and that the "cultural dimension" was seen as a "serious challenge." These quotations underline that the notion of culture is decisive to understanding this specific human-technology relation. However, the notion of culture, which is not developed further by the informants, seems very abstract; and, contrary to the predominant, more relativistic notions of culture, it indicates that, to a certain extent, culture is somehow fixed or stable. Yet if we are to take our informants seriously and try to understand the world they live in, we cannot simply brush them off with a relativistic remark that there is no such thing as a "Danish culture," suggesting that they are suffering from false consciousness.

We therefore suggest that the world(s) in which human-technology relations are played out is considered an "apparently stable world," since this is how the world is comprehended in many human-technology relations. While the world is not a specific world in Ihde's classic program 1, the *apparently stable world* is defined by factors such as nationality, history, cultural history, language (even dialects), nature, tradition, religion, economics, law, and ideology.[8] Such factors are generally taken for granted, but when a foreign technology such as Silbot is placed between the "I" and its familiar world, they suddenly—in the same way as the hammer does in Heidegger's trite, yet apt, example of the breakdown of ready-to-hand tools—re-appear as matters of concern.

Of course, it is impossible to write an exhaustive list of what an apparently stable world comprises. Nevertheless, the concept of an apparently stable world could help us understand with more sensitivity the technology that mediates between this world and the "I." Inspired by Jasanoff we understand the apparently stable world as a (often nationally) demarcated world of "historically contingent institutional foundations" (Jasanoff, 2012, p. 19) and of collectively held imaginaries of the future (Jasanoff & Kim, 2015). These cultural structures shape our different "civic epistemologies" (Jasanoff, 2005, p. 247)—that is, the way in which nations are collectively reasoning. Hence, to understand technology and its mediating role is necessarily also a question of understanding how the world, in different ways, can shape how the technology mediate and how the subject reasons. Furthermore, at an even deeper cultural layer, the apparently stable world is also a world of what Ardener (1982) calls *semantic densities*. In the "world of Finland" these might consist of such immanent and intuitively meaningful phenomena as Jean Sibelius, Kalevala, pines, cold winters, saunas, Uralic language, the Winter War, and, of course, the Moomins and mediating, while the world of pungmul music, calligraphy, mountains, monsoon, bathhouses, the Korean War, major hi-tech industries, and, not least, K-pop, is of course the "world of South Korea."

Studying the human-technology relations that occurred with Silbot would therefore most likely benefit from a detour into the different apparently stable worlds of Finland, Denmark, and South Korea. Theoretically stated, the expansion of postphenomenology's human-technology relations involves expanding the hermeneutic circle; that is, to see, listen, and tune into the spirit of the world in which the human-technology relations are played out and to understand the world as the cultural context that seems so essential to the embedding of technology culminating in technological style.

CLOSING REMARKS

In this chapter, we have aimed to grasp TOT within a postphenomenological framework by recognizing the complexity of the phenomenon and describing it as a two-way interaction. As such, the transferred technology, like the socially assistive robot Silbot, shapes the recipient culture, but the recipient culture also seems to shape the technology in the continuous effort to embed it in a recognizable, familiar praxis. Hence, the endeavor to understand the phenomenon of technology transfer calls for the cultural context of technology or external factors to be taken into consideration; in other words, it urges us to deal with the world. As we have claimed, in the well-known program 1, postphenomenology seems overly focused on comprehending the technological mediator that separates the human from the world. It takes the human "I" and the world for granted. We argue that the postphenomenological conception of both the human and the world need to be developed further.

To improve program 1, we introduced the notions of *the unstable human* and *the apparently stable world.* We suggested the "I" is always an unstable human and that a multistable technology mediates between the "I" and an apparently stable world. These notions can be seen as part of an effort to revive Ihde's cultural hermeneutics (1990) and to comprehend how culture shapes technology, or, put differently, how the world and the human mediated by technology become enmeshed in establishing the conventions of use of a given technological artifact. It is essential to consider these patterns of use that are characteristic of different cultural environments if postphenomenology strives to understand how the cultural context shapes the human-technology relationship and to pave the way for distinctive technological styles reflecting the apparently stable worlds.

NOTES

1. Most well-known is perhaps Jasanoff's works on biotechnology and genetic engineering (to mention just one example, see Jasanoff, 2005), but also notable is her comparison of nuclear-technology in the United States and South Korea (Jasanoff & Kim, 2009), where she and Sang-Hyun Kim introduce the by now rather influential concept of socio-technical imaginaries (see also Jasanoff & Kim, 2015).
2. This is the subject of Lasse Blond's forthcoming PhD dissertation to be published in 2018. Thus, some of the empirical examples are so far unpublished. Other authors have also written about the transfer of Silbot to Denmark (e.g., Hasse, 2013).
3. SILBOT was actually accompanied by another robot called MERO. In the shape of a bust, MERO could simulate different facial expressions, thus emphasizing Silbot's oral presentation of the cognitive exercises. However, due to technical problems, MERO was not used in Denmark, and, for comparative purposes, MERO will not be accounted for here.
4. "Robots in the Kustaankartano center for the elderly—The Living Lab pilot project and what was learnt from the INTRO Project" (2012) and "INTRO Interactive Robot Project." Final evaluation reports—unpublished.
5. "Final Evaluation Elderly Care Robot Systems in the City of Aarhus," (2012). Final evaluation report—unpublished. "Halvvejsrapport om Hjerne Fitness," (2011). Midway evaluation report—unpublished.
6. "Robots in the Kustaankartano center for the elderly—The Living Lab pilot project and what was learnt from the INTRO Project," (2012) and "INTRO Interactive Robot Project." Final evaluation reports—unpublished.
7. "Final Evaluation Elderly Care Robot Systems in the City of Aarhus," (2012). Final evaluation report—unpublished.

8. Like Hughes, we realize that these factors in writing become "overlapping, soft categories," (Hughes, 1986, p. 287).

REFERENCES

Ardener, E. (1982). Social anthropology, language and reality. In D. Parkin (Ed.), *Semantic Anthropology* (pp. 1–15). London: Academic Press.

Bloch, A. (2004). The Fouga airplane project. *Israel Studies, 9*(2), 1–33.

Gehlen, A. (1988). *Man, His Nature and Place in the World* (C. McMillan & K. Pillemer, Trans.). New York, NY: Columbia University Press. (Originally published 1940)

Hasse, C. (2013). Artefacts that talk: Mediating technologies as multistable signs and tools. *Subjectivity, 6*(1), 79–100.

Hughes, T. P. (1962). British industry lag: 1882–1888. *Technology and Culture, 3*(1), 27–44.

———. (1977). Regional technological style. In S. Strandh (Ed.), *Technology and Its Impact on Society: Symposium No. 1* (pp. 211–235). Stockholm: Tekniska Museet.

———. (1983). *Networks of Power—Electrification in Western Society, 1880–1930.* Baltimore, MD, & London: The Johns Hopkins University Press.

———. (1986). The seamless web: Technology, science, etcetera, etcetera. *Social Studies of Science, 16*(2), 281–292.

———. (1987). The evolution of large technological systems. In W. E. Bijker, T. P. Hughes, & T. Pinch (Eds.), *The Social Construction of Technological Systems* (pp. 51–83). Cambridge, MA & London: The MIT Press.

———. (1995). Shaped technology: An afterword. *Science in Context, 8*(2), 451–455.

Ihde, D. (1990). *Technology and the Lifeworld: From Garden to Earth.* Bloomington, IN: Indiana University Press.

———. (1993). *Postphenomenology—Essays in the Postmodern Context.* Evanston, IL: Northwestern University Press.

———. (2012). *Experimental Phenomenology (2nd ed.): Multistabilities.* Albany, NY: SUNY Press.

Jasanoff, S. (2005). *Design on Nature. Science and Democracy in Europe and United States.* Princeton, NJ: Princeton University Press.

———. (2012). *Science and Public Reason.* New York, NY: Routledge.

Jasanoff, S., & Kim, S.-H. (2009). Containing the atom: Sociotechnical imaginaries and nuclear regulation in the U.S. and South Korea. *Minerva, 47*(2), 119–146.

———. (Eds.). (2015). *Dreamscapes of Modernity: Sociotechnical Imaginaries and the Fabrication of Power.* Chicago, IL: Chicago University Press.

Lindqvist, S. (1984). *Technology on Trial—The Introduction of Steam Power Technology into Sweden, 1715–1736.* Stockholm: Almqvist & Wiksell International.

Pacey, A. (1990). *Technology in World Civilization—A Thousand-year History.* Oxford: Basil Blackwell.

Pantzar, M., & Shove, E. (2010). Understanding innovation in practice: A discussion of the production and re-production of Nordic Walking. *Technology Analysis & Strategic Management*, *22*(4), 447–461.

Plessner, H. (1981). Die Stufen der Organischen und der Mensch. Einleitung in die philosophische Anthropologie. In G. Dux, O. Marquard, & E. Ströker (Eds.), *Gesammelte Schriften* (vol. IV). Frankfurt: Suhrkamp Verlag. (Originally published 1928)

Rammert, W. (2002). The cultural shaping of technologies and the politics of technodiversity. In K. H. Sørensen & R. Williams (Eds.), *Shaping Technology, Guiding Policy* (pp. 173–197). Cheltenham, UK and Northampton, MA: Edward Elgar.

Rosenberg, N. (1976). *Perspectives on Technology*. Cambridge, London, New York, NY, & Melbourne: Cambridge University Press.

Ruttan, V. W., & Hayami, Y. (1973). Technology transfer and agricultural development. *Technology and Culture*, *14*(2/1), 119–151.

Scheler, M. (2009). *The Human Place in the Cosmos* (M. Frings, Trans.). Evanston, IL: Northwestern University Press. (Originally published 1928)

Shove, E., Watson, M., & Pantzar, M. (2012). *The Dynamics of Social Practice—Everyday Life and How it Changes*. London: SAGE.

Verbeek, P.-P. (2005). *What Things Do—Philosophical Reflections on Technology, Agency, and Design*. University Park, PA: Penn State University Press.

———. (2006). Materialzing morality: Design ethics and technological mediation. *Science, Technology & Human Values*, *31*(3), 361–280.

———. (2009). *Philosophy of Man and Technology. The Limits of Humanity: On Technology, Ethics, and Human Nature*. (Unpublished inaugural lecture). Faculty of Behavioural Sciences, University of Twente, Netherlands. (https://www.utwente.nl/bms/wijsb/organization/verbeek/oratie_eng.pdf)

Part IV

GENERAL METHODOLOGICAL ISSUES

Chapter 9

Why It Takes Both Postphenomenology and STS to Account for Technological Mediation

The Case of LOVE Park

Robert Rosenberger

Perhaps you've seen them. Small nubs, often metal, can be found affixed to horizontal surfaces in many urban spaces. They are called "skatestoppers," or "pig ears," among other colloquialisms. Their purpose is to discourage skateboarders from "grinding" their boards against these surfaces specifically, and to discourage them from skating in the area more generally.

There are several fascinating aspects of these common, low-tech devices. Despite their seeming innocuousness, they are a contested social maneuver. That is, although they are only small and simple protrusions, they represent a step in an ongoing dispute over the proper usage of public space. In addition, if you are not yourself a participant in this dispute—as perhaps a skateboarder, property owner, or some other stakeholder, advocate, or local government official—then it is possible to remain entirely unaware of these devices even as you make use of spaces that have them.

One thing that brings about this conflict between skateboarders and their antagonists is a particular skateboarding ethos, an aspect of skateboarding culture in which a special value is placed on exploring the streets and searching out new areas and surfaces that can be skated. In his book, *Skateboarding, Space, and the City*, Iain Borden (2001) analyzes this culture and concludes that skateboarding should be understood as a kind of performative critique of urban space. According to Borden, skateboarding is an "implicit

Figure 9.1 Skatestoppers built into a ledge in New York City, NY (Robert Rosenberger).

yet ongoing tendency to critique contemporary cities for their meanings and modes of operation, and to prefigure what a future unknown city might be. As one skateboarder declares, 'Skating is a continual search for the unknown'" (2001, p. 180).

In any case, when skateboarders make use of an urban space, it is often in discord with the plans for that space maintained by developers and owners. Sometimes those developers and owners have a different clientele in mind for their space. Sometimes they take exception to the damage to curbs and ledges that can be caused by grinding, or by the wax that skateboarders sometimes use to make such surfaces more suitable for grinding. In response, cities sometimes enact skateboarding bans, threatening skateboarders with fines and potential jail time for skateboarding in particular areas.

Such anti-skate laws often function in concordance with the skatestoppers and other anti-skate redesigns of physical space, like the skatestoppers. Skatestoppers are a paradigmatic example of what is variously called "hostile," "disciplinary," or "defensive" architecture—objects which control what people can and cannot do in certain spaces. A small skatestopper industry has even developed, with websites such as stopagrind.com and grindtoahalt.com.

But many skateboarders resist these attempts to close off spaces. They skate in defiance of bans. They develop new tricks that can work despite the existence of skatestoppers. And they even sometimes forcibly remove the skatestoppers themselves through the use of crowbars, hammers, and portable grinders or other power tools. The skatestopper manufacturers respond with the development of devices with greater tamper resistance. And the cold war

Figure 9.2 & 9.3 "No Skateboarding" signage and rail-mounted skatestoppers at Midtown MARTA Station, Atlanta, GA (Robert Rosenberger).

between these two social factions, built right into the architecture of public space, continues on.

It is my contention that despite their seeming simplicity, it requires the combined resources of social theory and the philosophy of technology to account for technologies like skatestoppers. In particular, I argue that insights from the field of science and technology studies (STS) are needed for capturing the ways that technologies are redesigned for the purpose of controlling specific social agendas. And I argue that we also need ideas from the philosophical tradition of phenomenology, and in particular the burgeoning school of thought called "postphenomenology," to conceive of the ways technologies are always open to multiple uses, even uses for which they were not designed. I will show that STS accounts of technology and postphenomenology are each incomplete—even in their own terms—without insights from the other.

As a plan for what follows, I begin with a review of some of the central ideas from postphenomenology and STS accounts of technology, and consider how these insights apply to the example of skatestoppers. Next, I review and expand arguments in favor of combining these two perspectives. Finally, I instantiate these ideas with the case of the history of Philadelphia's JFK Plaza, or "LOVE Park" as it is locally known, an area of the city's downtown that became an internationally renowned skateboarding destination, that is, until a strategic redesign and law enforcement crackdown largely forced the skaters out.

POSTPHENOMENOLOGY

As a blossoming international and interdisciplinary school of thought, postphenomenology provides and original account of the ways that technologies "mediate" human experience (e.g., Ihde, 1990, 1998, 2016; Verbeek, 2005, 2011; Hasse, 2015; Friis & Crease, 2015; Rosenberger & Verbeek, 2015). This focus sometimes earns this perspective the alternative moniker "mediation theory." Postphenomenology builds on classical phenomenology and American pragmatist philosophy to articulate deep descriptions of the various ways that human experience is transformed through technology usage. Rather than understand technologies to be like any of the other things we perceive or act on in the world, under this perspective technologies are conceived as transformative mediators of experience, coming between the user and the world, and changing the possibilities for perception and action. Peter-Paul Verbeek goes so far as to claim that users and the world itself become what they are through the mediation of technology (2011).

One of the key contributions of this school of thought is the list of different forms of human-technology relations developed by this perspective's founder, Don Ihde. Ihde explores the various ways that users engage with devices. He establishes a useful distinction between technologies that we take up with our bodies, and those we perceive directly and interpret. In what he calls an "embodiment relation" to technology, a device is brought into our bodily experience and transforms our capacities for perception and action (Ihde, 1990, p. 72). A pair of glasses, to review the default example, is not one of the objects I perceive in my environment, but something which transforms my visual capabilities. A hammer is not one of the objects of the world that I act upon, but something which transforms my capacity for action as it is embodied. This embodied form of human-technology relation can be contrasted with another, which Ihde calls "hermeneutic relations." This term refers to relations that involve looking at, listening to, or otherwise directly perceiving a device's readout (Ihde, 1990, p. 80). For example, a wall clock provides transformed access to the precise time of day when I look at and interpret its face. Hermeneutic relations are also at work when we interpret technologies that produce images, and a bustling line of postphenomenological research explores our relations to imaging technologies in everyday, medical, and scientific research contexts (e.g., Ihde, 1998; Hasse, 2008; Friis, 2012, 2015; Rosenberger, 2011, 2013; Forss, 2012; Carusi & Hoel, 2014; Wiltse, 2014; Vallor, 2015).

Ihde notes that our relations to technology at times become more or less "transparent." This refers to the degree to which a technology recedes into the background of our awareness, growing more and more unnoticed as it is used. Despite the fact that they sit perched on my nose, and despite the fact that they radically change my entire field of vision, I am so accustomed to my glasses that they often remain deeply transparent as they are worn. As I hammer nails into a piece of wood, the project I am attempting to accomplish is more present to me than my grip on the hammer itself.[1]

Several other forms of human-technology relations can also be identified. Ihde uses the term "alterity relations" to refer to technologies we interact with as if they have a "quasi-Other" presence, like the ATM machine with its question and answer format, or personal assistant smartphone applications (like the iPhone's Siri) that are voice interactive (Ihde, 1990, p. 98). Ihde also identifies "background relations" as those technologies that transform our environmental context while remaining off "to the side" in our experience, like a building's climate control system, or the gently humming refrigerator (Ihde, 1990, p. 109). More recently, Verbeek has tested the limits of this embodiment-to-background spectrum of relations. He suggests that technologies that are implanted into the body, such as a neurostimulator in the brain or a bionic eyeball, should be considered "fusion relations" since they problematize the very idea of *user, usage, embodiment*, and indeed the *human-technology relation* (Verbeek, 2011; Rosenberger & Verbeek, 2015, p. 21). On the other side, he identifies "immersion relations" in which the environment does not simply take on a backgrounded status, but instead actively shapes our experience, as in the case of "smart" homes (Verbeek, 2011; Rosenberger & Verbeek, 2015, p. 21).

Another central commitment of the postphenomenological perspective is that technologies always remain "multistable," that is, they have the capacity to mediate human relations to the world in multiple ways. As Cathrine Hasse explains, "In postphenomenology, multistability refers to technologies that vary in how their meanings are stabilized as they cross time. . . . Multistability emerges in the meeting of different kinds of practices in relation to the affordances offered by the design" (2015, p. 164). Any technology can be used in different ways for different purposes and can find different meanings in different contexts. At the same time, the notion of multistability also denotes the fact that technologies, as concrete material objects, do not simply enable just any mediated experience; a technology cannot be used for just any purpose, be meaningful in just any manner, or facilitate just any perception. The concrete materiality of the device places limitations on what kinds of relations can occur.

Even within the multiple possible uses and meanings of a given technology, there will often be a "dominant" one (or a few dominant ones), a main usage, usually the same usage for which the device has been designed and manufactured. When I wear my eyeglasses to see, I use them in their dominant stability. But other stabilities are possible, such as using them as a magnifying glass, or keeping an old pair for the sake of nostalgia.

Ihde uses the term "variational analysis" to denote the method of searching out the various stabilities possible for a given technology, be it through empirical investigation, armchair brainstorming, anthropological study, or other means. Ihde contrasts his own conception of variational analysis with the similar-though-not-identical version used by classical phenomenologist Edmund Husserl. If you want to appreciate postphenomenology's pragmatic commitments to anti-essentialism and nonfoundationalism, it is crucial to understand the difference between Ihde and Husserl on this methodological point. Ihde writes:

> Husserl's investigative method, patterned on mathematical variational analysis, was the use of what he called "imaginative variations," for which the result was supposed to be to determine *invariants* or *essences.* As argued, variational theory, in my estimation, is what gives phenomenology its rigor. But, again following Husserl, this time first in the first edition of *Experimental Phenomenology,* what I found was not a stable essence as Husserl also called his result, but multistability. (2016, p. 127)

That is, where Husserl's investigative method sought a target phenomenon's essence, postphenomenology seeks an appreciation of its multistabilty. As Shannon Vallor clarifies, "Ihde's concept of multistability undermines Husserl's original claim in *Ideas* to have founded a descriptive science of static eidetic essences; indeed, Ihde's explorations of human-technology relations have shown us that phenomena appear to us in far more fluid and open-ended ways than Husserl understood" (2015, pp. 19–20).

In recent work, I have expanded on the insights of Ihde's method of variational analysis, and developed a second step in this line of thinking for postphenomenology. My suggestion is that after variational analysis has been performed, and multiple stabilities have been identified, the investigation should continue through the critical contrast of those stabilities against one another. This enables postphenomenological methodology to do more than simply demonstrate the fact that a technology is multistable. Under this method, new things can be uncovered about the technology through the critical comparison of its various stabilities against one another. In this way, postphenomenological investigations can generate new insights without an appeal to either a foundational conception of truth or a transcendental account of essence; things are learned about *particular stabilities* through their comparison with *other stabilities*. This can be especially enlightening when investigating a technology's dominant stability, whose features can be occluded by its instantiation in the everyday lifeworld of many users. I refer to this second step for postphenomenological methodology as "variational cross-examination" (Rosenberger, 2014, 2017b; and, for an application to qualitative research methodology, see Aagaard, forthcoming).

We can consider the multistability of public-space devices in terms of skateboarding culture. As noted in the Borden quote above, a central aspect of this culture is the continual search through urban space for terrains available for skateboarding. Sidewalks and urban plazas become skate parks. Ledges and handrails become surfaces for grinding, as well as many other objects, from stone roadblocks to large planters. All of this is to say that through the lens of skateboarding we see many of the objects of public space to be multistable in a particular way: in addition to the designed and dominant usage of these objects of the urban landscape, they also afford stable possibilities for skating.

STS ACCOUNTS OF TECHNOLOGY

Some of the most influential thinking on the philosophy of technology over the past 30 years has come not from the philosophy of technology, but from sociologists and anthropologists in the larger and overlapping field of science and technology studies (STS). It is important here to consider some of the central ideas coming out of two theoretical perspectives from this field, actor-network theory and the social construction of technology.

The "social construction of technology" (SCOT) has been developed by Weibe Bijker, Trevor Pinch, and others (e.g., Pinch & Bijker, 1984; Bijker, 1995; Collins & Pinch, 1998; Pinch & Trocco, 2002; Oudshoorn, 2003). This view emerged as an offshoot of the "social studies of knowledge" (SSK) account of scientific practice. According to SSK, scientific knowledge develops through conflicts between research groups with competing social interests. In this account, focus is placed on the social factors involved in bringing closure to such conflicts. SCOT builds from these insights to develop a view of how technologies evolve through histories of disputes between collectives of people with competing interests within society. This view presents a powerful counter to any intuition that technology possesses an "internal logic" that determines how it should be constituted, and that technologies move forward only through the work of engineers and designers (Bijker, 2010, p. 71). Instead, in this view, the composition of any contemporary technology is the result of a social history in which one group's interests regarding how the technology should be used and composed won out over others.

SCOT researchers conduct detailed histories in order to discover what factors were important to the now-settled disputes that inform established technologies. A foundational study in this perspective is Bijker and Pinch's investigation of the history of bicycles (Pinch & Bijker, 1984; Bijker, 1995). According to this account, the shape of contemporary bicycles—with their two wheels of similar size—is not simply the result of the development of an ideal form independent of the context of concrete interests. Rather, the interests of those who wanted bicycles to be used primarily for travel won out over those who wanted them to be used for athletics and thus preferred a bicycle of a different shape (e.g., a bicycle with a very large front wheel). In the terminology of SCOT, any technology initially possesses an "interpretative flexibility," that is, a stage in which multiple designs coexist as debates occur over what that technology's form should take. Over time, social conflicts determine which particular purpose is the one a community prefers, and the technology develops along those lines (Pinch & Bijker, 1984, p. 421). Other examples of seminal case studies include Pinch's investigation of the development of electronic instrumentation, and Nelly Oudshoorn's history of oral contraception (Pinch & Trocco, 2002; Oudshoorn, 2003).

Another influential STS perspective is known as "actor-network theory" (e.g., Latour, 1992, 1999, 2005; Law & Hassard, 1999; Mol, 2002). Like the way that SCOT began as an application of insights from the SSK conception of scientific practice to technological development, the actor-network perspective also began as an account of science. It too conceives of scientific work in social terms, exploring the social factors involved in the closure of debates and the establishment of accepted facts within scientific communities. But in a crucial difference from SSK and SCOT, the actor-network account expands the kind of participants that should be analyzed as a part of social collectives. Under this view, such collectives should be conceived as "networks" that consist not only of human "actors," but also "nonhuman" actors, which in a scientific setting can include things such as laboratory space and instrumentation, patents, publications and their prestige, and affiliation with granting institutions.

The actor-network account of technology similarly looks at contemporary rivalries between social groups, and includes a conception of material objects as potential "actors" contributing to those networks. Under this view, technologies can be enrolled into networks to help enact the network's agenda. Human actors can do the work of "delegating" duties to nonhuman actors (Latour, 1999, p. 187). Bruno Latour, a leading light of the actor-network perspective, provides the example of the speed bump. We can imagine someone who would like to speed in their car down a particular roadway, and we can imagine this driver encountering a rival network of powerful actors that want that driver to slow down. These actors include the police department with its power to issue traffic citations as well as the state itself with its ability to determine traffic law. But in addition to the strategy of distributing tickets to those who are caught breaking the speed limit, the slow driving network has another option available to it: the task of forcing the driver to drive slowly can be delegated to a speed bump. In this way the speed bump, as a material actor (through the work of the engineers and construction workers that design it and put it in place), does the work of the police officer. Latour writes, "Not only has one meaning, in the example of the speed bump, been displaced into another, but an action (the enforcement of speed law) has been translated into another kind of expression. The engineer's program is delegated in concrete" (1999, p. 187). And he notes that in many countries speed bumps are referred to as "sleeping police officers." The work of those legislators, engineers, construction workers, and others all confront the speeding driver in the form of the bump. As Latour summarizes, "Think of technology as congealed labor" (1999, p. 189).

Madeleine Akrich contributes to this line of thinking by describing the kind of work that must be done to a material object to enroll it into a particular network (1992). To describe the agenda of a network, she makes a metaphor to the "script" of a movie or a play, the narrative according to which the various actors follow a role. We can thus think about the material modifications that are made to technologies to better align them with their roles in the network as a kind of "inscription," inscribing the script of the network into the device to make it a better actor. She writes, "A large part of the work of innovators is that of '*inscribing*' this vision of (or prediction about) about the world in the technical content of the new object" (Akrich, 1992, p. 208). In the case of the speed bump, we could potentially even consider the bump itself as an inscription into the roadway, further enrolling the street into the slow speed network. We could think of additional material changes, such as painting the bump so that it is more visible, or planting a sign beside the bump to further warn drivers, as additional material inscriptions.

In my view, STS accounts of technology in general, and actor-network theory in particular, are particularly well positioned to account for the social dynamics of skatestoppers. We can conceive of the skateboarders and those who oppose the skateboarders as rival networks. Under this account, the various potentially skateable objects of public space can be understood as material nonhumans that each network attempts to enroll. Thus, the skateboarding network enrolls the various sidewalks, plazas, handrails, benches, ledges, and other skateable spaces and objects into its network. The wax that skaters sometimes apply to ledges to render them more grindable can be conceived as a material inscription. And thus the skatestoppers themselves can be conceived as material inscriptions as well, but ones implemented by the anti-skate network. The skatestoppers are modifications to those ledges and handrails to which are delegated the task of discouraging skateboarding.

The postphenomenological method of variational cross-examination adopts some of the ideas from actor-network theory (Rosenberger, 2014; 2017b). Recall that this method involves first identifying different stabilities, and then cross comparing them. One of the features of the various stabilities that can be cross compared are the different ways that each stability may have been materially inscribed to better fit its usage. This can be important if we do not already know the functions of all of the parts of our object of study. And it can be especially important when the dominant stability of a device is under examination. As noted above, it can be difficult to ascertain many of the features of a dominant stability since it may be suffused in normality, set in the habits of the everyday world. This is even more the case if a powerful network has further tailored the device to the specifications of this dominant stability. It can become even easier to believe that the device is only "for" this dominant purpose.

Think again of the skatestoppers. If one is neither a skateboarder nor familiar with the debate over skateboarding in public space, then one might not even notice skatestoppers in one's everyday environment. And even if one does, one may not guess their purpose since the device may have effectively pushed away the skateboarders (and since the dominant anti-skate network may have also additionally effectively pushed them away through other means, such as anti-skate laws).

ON THE DIFFERENCE BETWEEN THE NOTIONS OF INTERPRETIVE FLEXIBILITY AND MULTISTABILITY

At this point a clarification should be made: the postphenomenological notion of "multistability" and the SCOT notion of "interpretive flexability" should not be understood as entirely interchangeable. Because they point to similar phenomena—the general variability of technology usage and meaning—these two concepts are sometimes mistakenly understood to point to the exact same thing. It appears that even Verbeek conflates the two when he writes that "the existence of multistability—a product's 'interpretative flexibility' as Bijker (1995, p. 20) calls it—need not hamper designers in explicitly trying to anticipate the mediating role of products in their use context" (Verbeek, 2005, p. 217).

My insistence here to keep the notions of multistability and interpretive flexibility distinct is no mere quibble. Not only does each concept point to a subtly different phenomenon, they both are individually important. Where the notion of interpretive flexibility refers to a technology's status in a social community, the notion of multistability refers to a quality always inherent in all technology.

Let's spell this out further. Interpretive flexibility is in the end a social concept. It points to a technology's particular relationship with a larger community of human users. As Bijker puts it, "Because the description of an artefact through the eyes of different relevant social groups produces different descriptions—and thus different artefacts—this results in the researcher's demonstrating the 'interpretative flexibility' of the artefact" (2010, p. 23). A technology is understood to have interpretive flexibility when its most basic aspects, its very definition, its default form, its agreed up function, are up for debate. This can occur in the early stages of a technology's development. We see this in the example of the bicycle. If a group of people were asked today to imagine a normal bicycle, it is most likely that most people would think of the same image, one in which the bicycle's design has wheels the same size. Sure, any of these people will also recognize there to be any number of other bicycle designs out there, from tiny-wheeled models, to three-wheeled models, and more. But these alternative bicycle designs will be recognized for what they are: atypical ones. Thus today the technology of the bicycle does not possess much interpretive flexibility. Socially speaking, the basic definition of the bicycle, its standard and nonstandard designs, and its normal and atypical uses, are generally agreed upon. Indeed, to even remember a time in which this was not the case requires the work of opening the blackbox of history on this topic, as Pinch and Bijker have done.

As we have seen, when that historical investigation is performed, it is revealed that in an earlier time the term "bicycle" referred to things with very different basic designs and uses. If a group of people were polled at that earlier time to determine what the term "bicycle" meant, then more than one agreed-upon answer would have been given. That is, in the early chapters of the story of the bicycle, there was interpretive flexibility.

It is of course not only in the early stages of a technology's place in a community that interpretive flexibility can occur. As changes occur over the course of technological development, and as a community changes over time, a previously settled technology can once again become interpretively flexible. This appears to be the case for our contemporary relationship to the computer. In the 1980s and 1990s, if you asked a group of people to imagine a computer, then most answers would have pointed to the same thing: a desktop model with a keyboard, mouse, and monitor. Today it is quite possible that the same question would prompt a variety of answers, including laptops, tablets, smartphones, wearables, as well as desktops. In this way, we could say that there is currently interpretive flexibility with regards to the computer.

Perhaps adding to the confusion between these concepts is the fact that both postphenomenology and SCOT differently use the term "stable." Where postphenomenology refers to the "stabilites" of a multistable technology, STS accounts refer to the "stabilization" of a technology with regard to interpretive flexibility. In SCOT terminology, as a technology becomes less interpretively flexible, as its basic design and purpose becomes settled within the community, the technology is said to "stabilize." As Bijker explains, "'Stabilization' stresses the process character: a process of social construction can take several years in which the degree of stabilisation slowly increases up to the moment of closure" (2010, p. 69). Again, like interpretive flexibility itself, for SCOT the notion of stabilization is a social concept, referring to the social state of a community of people with respect to a technology.

In contrast, the notion of multistability points to an always-present condition of all technology. It points out the pragmatic nature of the relationality between material devices and users. In its simple but important claim that the materiality of a technology always makes possible multiple relations to the world—while simultaneously also always constraining possible relations—the notion of multistability constitutes a kind of metaphysics. To say that technologies are always multistable is to make an ontological claim, but an odd one. It's odd in that it is an ontology that is nonetheless non-essentializing and non-foundational and non-deterministic. It is a metaphysical claim about the always-present pragmatic relationality between users and technologies.

THREE ARGUMENTS FOR AMALGAMATION

The basic observation behind the push to bring together postphenomenology and STS accounts of technology is that each perspective appears to possess strengths that the other lacks. They have complimentary specializations. In the view of those who favor putting together these two approaches, the STS accounts capture the ways that technologies are shaped by social collectives, and also the ways those technologies themselves contribute to the ongoing dynamics of those collectives. The postphenomenological account instead captures the relations between individual users and technologies. In this section, I not only support amalgamation, but argue for a strong version. This strong case for amalgamation, as I identify it here, includes not only stressing why it would be beneficial to combine these theories, but also arguing that each of these perspectives is incomplete even in its own terms without the integration of insights from the other. That is, I argue that without the insights of the other, each of these perspectives are inadequate even for approaching its own main object of specialization.

Let's consider three arguments in favor of amalgamation. The first, which can be called the "argument from interface," claims that postphenomenological insights are important to STS accounts for the way they articulate the details of human-technology relations. Versions of this argument have been advanced by Peter-Paul Verbeek and Aaron Smith, and I build on their insights here. The second, which can be called the "argument from multistability," is my own. I suggest that postphenomenological insights are additionally crucial to STS accounts for the way they articulate the nature of technological materiality and its inherent openness to multiple uses and meanings. The third and final argument, which can be called the "argument from inscription," spells out the way that postphenomenology itself requires ideas from STS accounts of technology, and in particular its insights into the ways that technologies are materially inscribed into social networks. Each of these arguments is clarified through its application to aspects of the example of skatestoppers.

The Argument from Interface

When postphenomenologists have urged for the importance of the insights of their perspective for STS work on technology, they have tended to make this case in terms of the following reasoning.

First, it is understood generally that postphenomenology shares many philosophical commitments with STS accounts of technology, and with actor-network theory in particular. They share, for example, a commitment to a relational ontology, that is, the basic idea that things and people can only be understood in terms of their relations to other things and other people. Or as Donna Haraway puts it, "beings do not preexist their relatings" (2003, p. 6).

From here, the argument goes that while STS accounts of are good at spelling out the relations between groupings of people and things, and how these groupings work together to enact agendas, these accounts are not as strong when it comes to articulating the details of the relations between individual humans and devices. Thus, according to this line of thinking, STS accounts and postphenomenology should be combined because postphenomenology fills in this gap; it is good at articulating the details of individual human-technology relations. Put differently, postphenomenology and STS should be combined because the former crucially supplements the latter with an account of the "interface" between individual humans and technologies. Let's refer to this general line of reasoning here as "the argument from interface."

The two fullest articulations of this argument have appeared in the work of Peter-Paul Verbeek and of Aaron Smith. In a chapter in *Chasing Technoscience*, a book which cross compares several different philosophical approaches to the study of materiality, Smith offers a clear formulation of this reasoning, writing that, "Latour's view, however, does not develop in nearly the same depth the direct personal relationships with artifacts that Ihde's does. Instead, Latour's project could be seen as picking up where Ihde's ends because it emphasizes systems of relations" (Smith, 2003, p. 189). But it is in the work of Verbeek (2005, 2011) that this argument receives an extended treatment. He writes, "Actor-network theory is primarily interested in unraveling the networks of relations by virtue of which entities emerge into presence, while a postphenomenological approach, by contrast, seeks to understand the relations that humans have with those entities" (Verbeek, 2005, p. 164f).

One way that both Verbeek and Smith articulate this view is in terms of Latour's comments on gun usage. Latour cleverly clarifies the importance of a relational conception of ontology through the analysis of the rhetoric around gun control. He notes that gun rights defenders like the National Rifle Association (NRA), a powerful anti-gun control lobbying group in the United States, often point out that "guns don't kill people; people kill people." The claim of these lobbyists is that responsibility for shootings must always rest only on a person, thus as a community we should not place limitations on access to guns. Latour spells out this logic: "The gun is a tool, a medium, a neutral character of human will" (1999, p. 196). This reveals how the shallow rhetoric of the NRA works to obscure the issue. As Verbeek further explains:

> A gun is not a mere instrument, a medium of free will of human beings; it helps to define situations and agents because it offers specific possibilities for action. A gun constitutes the person holding the gun as a potential killer and his or her adversary as a potential lethal victim. (2011, p. 64)

Or as Latour puts it:

> You are a different person with the gun in your hand. (1999, p. 179)

But the example of gun usage does more than only helpfully instantiate actor-network and postphenomenological concepts. It also serves as an example of the importance of postphenomenological insights for actor-network theory's account of technology. Verbeek writes that, while Latour's framework enables us to identify chains of actors, "the postphenomenological perspective, for instance, offers a more nuanced look the connections between the entities in its chains. Latour views these connections simply as 'associations,' as a kind of cement between actants" (2005, p. 165). And Verbeek notes that this nuance is lacking even in Latour's iconic example of the gun, since in Latour's account gun usage reduces to a description in which a new entity emerges as "gun + human = gunman." Where Latour's account can only get us to the understanding that a user and a technology come together to form a new networked entity with new capacities, postphenomenology has the potential to spell out further how this happens. It does so with its conceptions of the human embodiment of technology, hermeneutic interpretation and perception, and the experiential transparency involved in bodily habituation and training.

I'd add as well that Latour's conceptual vocabulary is not neutral with regard to the description of individual human experience within networks; it actively obscures certain dynamics (Rosenberger, 2014). Latour's project, at least in his pioneering mid-nineties work on technology, is to draw out the sociality of objects, and also the sociality of human bodies. That's why he uses the term "delegation" (a word usually used only to describe relations between human bosses and human subordinates) to denote the assignment of a task to a device. Conversely, he uses the term "incorporation" (a word usually used to describe something done to objects) to denote the enrollment of a human into a network (Latour, 1992, p. 231). While these moves are helpful for Latour's particular project, rhetorically emphasizing the transfer of "competences" between humans and nonhumans, I suggest that they have the effect of actively obscuring (and perhaps also downplaying) issues of individual human experience. Latour is sometimes criticized on the grounds that his conceptual framework has a leveling effect, reducing important differences into a flat ontology in which everything is just another actor (e.g., Lee & Brown, 1994). The argument from interface, exemplified by the gunman critique above, is a particular version of this criticism of actor-network theory's flattened ontology. While it is at times important to emphasize the larger sociality of objects and human bodies, this should not always be done at the expense of a nuanced account of human experience itself, the details of which are at times crucial to understanding the dynamics of a network. Postphenomenology can provide a valuable corrective on this point.

The discord between the ultimate project at hand for STS accounts of technology and postphenomenology thus runs so deep that it is fixed within the connotations of the very vocabulary of each perspective. This underscores an important point: just because these perspectives happen to specialize in the articulation of things the other lacks, that should not imply that the amalgamation of the two will be an easy matter. Indeed, just the opposite appears to be the case. Because the lenses offered by each perspective have been grinded to enable a focused view upon different aspects of the world, it may be difficult, or even perhaps impossible, to entirely reconcile the two such that a complete amalgamated account can be taken up together in a given moment.

These points crystalize in a simple example: rumble strips. Recall that in the case of Latour's example of the speed bump we get a simple instance in which the capacities of the state's police institution are transferred to a particular device. The state's agenda of forcing drivers to slow down is delegated to the speed bump. Rumble strips offer a related and equally simple example, but with different processes at work. By "rumble strip," I refer to the practice of cutting a series of grooves into the road that produce a rumbling sound and tactile sensation when the car rides over them (or installing mini-bumps that create a similar effect). These kinds of grooves are sometimes installed within roadways to alert drivers to upcoming changes, for example, issuing a reminder that a toll both is ahead. They also are sometimes added to the sides of the road to alert drivers who have veered too far out of their lane.

Although they are built into the roadway, the rumble strips play a slightly different material role from the speed bump. Where the speed bump coerces the driver by force, threatening to damage a car's suspension like the way a police officer threatens the driver with a fine, the rumble strips instead only supply information. In this way, they are functionally more like a street sign, but the form of their mediation is different. In the case of visual signage, spelled out in words and symbols, a sign's content is apprehended through a visual hermeneutic relation to technology. If the driver already knows how to read those words and symbols, then she or he experiences their meaning in a visual gestalt. The rumble strip also relays a signal to the driver through a hermeneutic relation, just like the visual signage, but a tactile one. And this haptic hermeneutic relation occurs through the context of a driver already deeply embodying the car.

There's more. Depending on the particular driver's experience with individual instances of particular rumble strips, their tactile perception may come as more or less of a surprise. A driver already deeply familiar with a particular rumble strip (say, one passed over as part of a daily commute) may experience it with some degree of transparency, anticipating the tactile sensation before it occurs, already aware of the information it transmits. A driver unaccustomed to a particular strip, or otherwise unready for its sensation, may experience it as a jarring alert which pierces into their overall awareness.

Thus, although in a certain sense very similar to the speed bump (i.e., both are built into the road in order to influence the behavior of drivers that ride over them), the example of the rumble strip entails many distinct details usefully captured by postphenomenology. And these are details that actor-network theory alone may actively obscure, accounting for only another "association" between an "incorporated" driver, a car, and a roadway alteration, themselves all actors.

The argument from interface is relevant to the example of skatestoppers. It perhaps goes without saying that skateboarding is an activity that requires finely developed bodily skills. This is the case not only for everyday skating, and for performing tricks off of common objects like railings, ledges, and roadblocks. It is also true for those who develop the ability to skate across surfaces studded with skatestoppers, and even also for those who become proficient in skatestopper removal. Put into the terminology of actor-network theory, postphenomenological insights are useful for describing the nuances of the bodily incorporation involved in the enrollment of contested public-space actors into their network. And more, there is a hermeneutic skill involved in the search through public space for zones suitable for skating. Part of the perceptual hermeneutics of skateboarding involves the bodily perceptual training involved in seeing and feeling the environment to afford new possibilities for skateboarding.

Thus, while STS accounts are perhaps ideal for capturing rivalries between these social collectives (collectives that can include material actors), postphenomenological insights are useful for articulating the very different ways that human users within one network relate to contested devices differently than users from the rival network.

The Argument from Multistability

I argue that STS accounts of technology are incomplete, even in their own terms, without a conception of multistability. My suggestion is that while STS accounts of technology are ideally situated to capture many of the ways that social rivalry shapes technology, and the parts technologies can play within those rivalries, these accounts still require an additional philosophical conception of materiality if they are to apprehend the reasons why particular groups come into conflict over particular technologies in the first place. And I suggest that the notion of multistability is well suited to the task.

The notion of multistability is useful for conceiving of the ways that technologies, as material artifacts, are at the same time both open to multiple uses and yet also not open to just any usage. As noted above, this means that multistability is not an exclusively social concept. In this way, the notion of multistability does provide an ontological conception of technology, but a non-traditional one since it does not point to transcendental essences. It is instead a practical concept emphasizing any object's limited-though-multiple potential fit into human ends and interpretations, in tune with postphenomenology's commitments to the tenets of American pragmatist philosophy. This is all to say that the reason that particular material artifacts have the potential to afford different uses and meanings to different social groups is exactly because technology is multistable.

The example of skatestoppers is a case in point. STS accounts of technology provide an ideal means for following out the social conflict over public-space objects between skaters and those looking to deter skateboarding. But an ontological conception of materiality in general, and the notion of multistability in particular, is useful for explaining how particular devices draw these two particular groups into conflict in the first place. The technology's role in drawing these two groups together is non-determinative; it didn't force them into conflict. But it's openness to supporting both the skating and non-skating relations was one of the conflict's necessary conditions.

For example, we can imagine a story in which a particular alleyway has not historically been a special draw to skateboarders. The network of the private alleyway owner and the network of local skaters were not in conflict. Then imagine that a new staircase is installed in that alley, and it includes a handrail that happens to afford skateboard grinding in a way that raises the interest of the local skater community, and let's imagine that in this story the alleyway owner is displeased with these events. In this way, the very particular multistability of the device itself is part of what makes the conflict between these two networks (and not just any two networks) possible. Sure, the railing may then be retailored with skatestoppers to enroll it further into the anti-skate alleyway owner's network, and thus disenroll it from the skaters' network. And the tools of actor-network theory would be well suited for spelling out those dynamics. But the original multistability of the handrail itself played an indelible agential role in bringing about the rivalry of these two networks in the first place.

The Argument from Inscription

I argue here that the postphenomenological account of technology is incomplete, even in its own terms, without incorporation of insights from STS. My suggestion is that while postphenomenology is ideally situated for the description of individual human relations to technology, this perspective alone cannot account for the ways that the technology has been altered by others for the purpose of changing these very relations. STS accounts are well positioned to capture much of these additional relevant dynamics. Thus, for postphenomenology to more completely accomplish the project of its own specialization, it will be helpful to adopt ideas from STS accounts of technology.

As a start, Akrich's notion of "inscription" is, in my view, one that should be of particular relevance to postphenomenology. As incorporated into actor-network theory, the notion of inscription refers to the material changes made to a technology so that it better fits into the "script" of a particular network. Such changes enable a device to better fulfill its role within that network. When looked at in terms of the phenomenological experience of individual users, such inscriptions can be understood to facilitate a particular relation set in terms of a network's particular social script. That is, the particular quality of a user's experience, and the purpose most readily afforded for that user by the device, are in part determined by a device's material inscriptions, engraved into the device through the work of powerful actors.

The effects of material inscriptions upon user experience can be most stark when they work not only to facilitate a particular usage, but to also make it difficult or impossible to use that device for a particular alternative usage. I suggest that here is a place where postphenomenology and STS accounts of technology can be combined in a technical and potentially fruitful way: multistable technologies can be reconceived as maintaining various stabilities that may be facilitated or shut down by social networks through material inscription. It is possible to develop a theoretical framework to spell out the ways that social groups "close off" or "reopen" or otherwise restrict or broaden the particular stabilities that a technology enables (Rosenberger, 2017a, 2017b). That is, we can look into the ways a network utilizes material inscriptions to "close off" a particular stability of a technology, rendering it difficult or impossible to use for some undesired purpose. We can also look into the ways a rival network may engage in a counter-inscriptive process, and "reopen" that formerly closed stability. We could even look into inscriptive strategies that work to facilitate multiple stabilities, that "open up" multiple usages, rather than close them off.[2]

Skatestoppers are a prime example of a material inscription that can be understood in these terms. Public-space objects like handrails and ledges are multistable, open to both their dominant stability and also the alternative skateboarding stability. The skatestoppers are material inscriptions that work to close off the skateboarding stability. When skaters return the favor and forcibly remove the skatestoppers with crowbars or power tools, this can be understood as an act which "reopens" those surfaces. These acts of removal, these instances of vandalism, work to reinstate a stability formerly closed off by the skatestoppers.[3]

Whether we should favor inscriptions that work to "close off" or "reopen" a particular technology will depend on our values, agendas, and perspectives on a particular case. None of this work is inherently better or even more progressive. It depends on what is actually happening in each instance.

THE CASE OF LOVE PARK

As I write these words, an area known as JFK Plaza in Philadelphia, PA, also nicknamed "LOVE Park," is undergoing a radical renovation. This construction project can be understood as a coda to a decades-long drama about a struggle over the proper use of a particular plot of public land, but also, more largely, over the identity of a city. This struggle has been waged through the enactment of laws, police action, community activism, large-scale protests, and civil disobedience. It was also crucially waged through the strategic redesign of public-space objects.

Built in the 1960s, JFK Plaza sits in Center City Philadelphia at the terminus of the Ben Franklin Parkway, across from City Hall, and just a block from Broad Street. It acquired the name "LOVE Park" as a reference to Robert Indiana's iconic statue of the word "love" (in big red capital letters with the

Figure 9.4 LOVE Park, Philadelphia, PA, USA (Robert Rosenberger).

"O" askew), first added to the plaza for the bicentennial, then acquired permanently in 1978.

As recalled by Edmund Bacon, the urban planner responsible for the plaza, he and his colleagues "never would have had the slightest premonition that our work would become world famous, and I think that if we had tried to do something world famous we would have missed" (411 Productions, 5:25). But that is exactly what happened. In the 1980s and 90s LOVE Park became a major location for skateboarders, both as a training ground for local talent, and as a destination for visitors from around the country and around the world. The plaza was featured on magazine covers, video games, and skateboarding industry advertising. Due in part to the visibility brought by LOVE Park, Philadelphia was chosen in 2002 and then again in 2003 as the site for the X Games, the nationally televised alternative sports competition.

Despite the fact that the planners and architects did not design LOVE Park to be a skatepark, the space has proven multistable in just these terms. As Philadelphia skater Josh Kalis explains at the start of the documentary *LOVE Story: The Saga of a Skate Landmark*, "LOVE is so perfect, and it's set up so right, that, you know, you can't get any better than that place" (411 Productions, 0:12).

A variety of individual elements throughout the plaza afforded a stable skateboarding stability. Skaters would grind along ledges, steps, benches, railings, and roadblocks throughout the space. Occasionally someone would

Figure 9.5 The steps leading down into the LOVE Park fountain (Robert Rosenberger).

"pop" a large ground tile, prop up one side, and use it as a makeshift jumping ramp. The fountain, when empty, was a centerpiece of skating events, as especially skilled skaters could jump the "gap," launching over the steps and down into the fountain bed.

By the turn of the century a powerful selection of government officials decided that they had seen enough of the skateboarders' antics, and worked to shut down skateboarding in LOVE Park. Stated reasons for this push included liability issues, and the cost of damage to the plaza from skateboarding. In his study of these issues, Jeremy Németh questions whether these were legitimate concerns, pointing out that the liability of the plaza should be no more than other parks where sports are played, and that the damage costs were far lower than the millions brought to the city through the X Games (and far lower than the cost of the city's later renovations), and also that private skateboarding-related companies volunteered to pay for the damages (2006, p. 301).

In place of these stated reasons, we can look at the issue of skateboarding in terms of disagreements about the future of downtown Philadelphia. On the one hand, some saw value in the youthful energy and profile that the LOVE Park skaters brought to the city, with bustling skate-related businesses, as well as the presence of the X Games as an example of the positive national attention accrued. Architectural historian Ocean Howell argues exactly this: "Skateboarding may have been a wrench in the modernist 'growth machine,' but in Philadelphia, it has been retooled as a cog in the entertainment machine" (2005, p. 41).[4] On the other hand, others saw the skaters to run against the image they desired for Center City, that is, that of a professional business and tourist district. As Németh concludes, although it is difficult to "uncover the City's genuine rationale for their actions in the face of unrelenting popular support for the skateboarders," it can be reasonably speculated that "the skateboarders' 'polluting presence' in LOVE Park did not support the City's desired image for their redevelopment efforts" (2006, p. 315).

In an effort led by then-councilman (and later, Mayor) Michael Nutter, the Philadelphia City Council passed a Municipal Code in 2000 that banned skateboarding in LOVE Park. Then in 2002, in an effort championed by then-mayor John Street, the city embarked upon an $800,000 renovation of the plaza. As Rick Valenzuela wrote at the time for the *City Paper*, "It's doubtless that history will remember Mayor John Street as the hangman who killed Love Park" (2002).

In his pan of the redesign after its unveiling, Inga Saffron, the architecture critic of the *Philadelphia Inquirer*, explains, "To block the skateboarders' path, they installed an array of sickly pink concrete planters and trash cans," and notes that, "It may be the first time in the history of park design that trash cans are a dominant visual feature" (2002, p. D01). As Németh notes, the renovation included replacing the skateable granite benches with "wooden benches with crossbar dividers around the park to serve the dual purpose of deterring both skateboarders and reclining homeless persons" (2006, p. 301).

The skateboarding community pushed back against these changes through a variety of means. Skaters continued to make use of the plaza, despite occasionally getting chased by cops, and despite the redesign (e.g., see the ledges in Figures 9.6 and 9.7, darkened through grinding and waxing). Large rallies were held in protest of LOVE Park's closure for renovation. At one rally, a then-92-year-old Edmund Bacon defiantly (and adorably) rode across the plaza on a skateboard and shouted that he, "in total defiance of Mayor Street

Figures 9.6 & 9.7 Images of ledges in LOVE Park after the 2003 renovations, with waxed ledges, and strategically placed planters (Robert Rosenberger).

Figure 9.8 A bench in LOVE Park after the redesign (Robert Rosenberger).

and the Council of Philadelphia, hereby exercise my rights as a citizen of the United States. I deliberately skate in my beloved LOVE Park" (McQuade, 2016b). An advocacy group developed an alternative design for the renovation, and worked with the skating industry to raise a huge sum of money for renovation and continual upkeep the plaza. All of it ultimately came to little avail.

Illegal skateboarding continued in LOVE Park until early 2016 when it was closed down once again, this time for a much more radical renovation that will add new green spaces, and at least from the appearance of the plans, functionally cut off skateboarding in the park once and for all. As a nod to its historic importance to the city, skateboarding was once again made legal for five days before the area was closed off for construction (McQuade, 2016a). New design specs for the park make it seem like it will be even less skateable than it was even after the 2003 renovations.

The case of LOVE Park instantiates many of the concepts reviewed above, and in extreme and instructive ways. We see an example of multistability in which an entire area was designed and built for one purpose and became world renowned for another purpose entirely. In this same case we see a powerful network, including the mayor and part of the City Council, to work against those who had taken up this area of the city for this second purpose, and to decisively shut down the alterative usage through material inscriptions in the form of large-scale renovations.

CONCLUSION

There are important political and ethical moments that occur at the junctures between collective agendas and individual actors. They occur as individual humans or technologies take part in the plots of those larger groups. They occur as these people or devices run up against an agenda that moves in a direction counter to their own. And they occur as those people or things sit at the tense intersection between rival networks, and become, themselves, the fought-over sites of that rivalry. I suggest that these crucial political and ethical moments are best captured through a strategic combination of insights from postphenomenological theory and STS accounts of technology. And I suggest that these moments are best uncovered through the methodology of variational cross-examination.

With the example of skatestoppers we see two important things at once: the multistability of a variety of simple public-space technologies as viewed through their alternative skateboarding stability, and a widely used material inscription that targets that skateboarding stability. With the case of LOVE Park we see these things in a specific and large-scale manner: the multistability of a plaza that while not built for skateboarding nevertheless became a skateboarding institution, and the forceful and effective effort to terminate this alternative stability by a powerful rival network, namely the city government and its affiliated business interests.

In closing, it should be noted that while I do hold that this combination of postphenomenological and STS insights has the potential to reveal crucial moments of politics and ethics in a way that perhaps no other theories can manage, I do not mean to imply (as others sometimes have) that these combined theories have any potential to provide a complete account of politics or ethics.

They do not, for example, maintain accounts of political structures like democracy, economic systems, rights, penalty, or difference. They do not provide complete guidance on how to address specific ethical questions, or a general account of the good life. However, I do suggest that the combined insights of these theories do offer unique and essential contributions to the larger discussions on ethics and politics that may not be available anywhere else.

NOTES

1. Postphenomenology is at once deeply and expansively indebted to the work of Martin Heidegger, and at the same time an attempt to move beyond what is sometimes considered the foundation-seeking nature of his work in its attempt to categorize the modes of being. In particular, Ihde's notions of "embodiment relations" and "transparency" straightforwardly borrow from Heidegger's account of tool use (to which the hammer is a reference), and also the work of Merleau-Ponty which resituates these ideas within an account of the body (Heidegger, 1996; Merleau-Ponty, 1962).

2. For example, I have explored the multiple stabilities afforded by public-space devices such as benches, trash cans, and fire hydrants. In addition to their role in fighting fires, hydrants are also sometimes opened as a way to keep cool during the summertime heat, especially within economically disadvantaged communities. Cities will sometimes close off this stability through the use of hydrant locks (Rosenberger, 2017b). The homeless are a population often targeted by hostile architecture designs. In conjunction with wide-ranging policies on loitering, panhandling, and camping, cities sometimes utilize dividers and armrests that cutoff a bench's use as a bed, and rainhoods and locks that cutoff a trash can's use as a source of food or recyclables, among many other hostile designs, all working together to force homeless people out of public spaces and to render the problem of homelessness less visible (Rosenberger, 2014, 2017a). See also my for-the-public writing on the topic, for example: www.theatlantic.com/business/archive/2014/06/how-cities-use-design-to-drive-homeless-people-away/373067/

3. Snøhetta, the designers of the Oslo Opera House, also known as the Norwegian National Opera and Ballet, consulted with skateboarders to learn how to include outdoor surfaces that facilitate skating (Blum, 2008). This is in an example of work that "opens up" stabilities. It reinforces the potential to approach the space through a variety of stabilities.

4. Howell contrasts this view with the one attributed to Borden (2001) (who is quoted earlier in this chapter), where Borden suggests that skateboarding can be conceived as a form of critique, one that analyzes and exposes aspects of the city without contributing to it. Howell (following the influential work of Ricard Florida) holds that skateboarders should instead be understood as part of the "creative class," drivers of the economy through youthful image, artistic advance, and the economic benefits to a city that come follow those virtues (2005).

REFERENCES

411 Productions. (2004). *Love Story: The Saga of a Skate Landmark. ON Video Magazine*, Winter issue. https://vimeo.com/19819372

Aagaard, J. (2016). Introducing postphenomenological research: A brief and selective sketch of phenomenological research methods. *International Journal of Qualitative Studies in Education.* http://dx.doi.org/10.1080/09518398.2016.1263884

Akrich, M. (1992). The de-scription of technical artifacts. In W. E. Bijker & J. Law (Eds.), *Shaping Technology/Building Society* (pp. 205–224). Cambridge, MA: MIT Press.

Bijker, W. E. (1995). *Of Bicycles, Bakelites, and Bulbs: Toward a Theory of Sociotechnical Change*. Cambridge, MA: MIT Press.

———. (2010). How Is technology made?—That is the question! *Cambridge Journal of Economics, 34,* 63–76.

Blum, A. (2008, November 24). New Oslo opera house is really a stealth skate park." *Wired.* http://www.wired.com/2008/11/pl-design-7/

Borden, I. (2001). *Skateboarding, Space and the City*. London: Berg.

Carusi, A., & Hoel, A. S. (2014). Toward a new ontology of scientific vision." In C. Coopermans, J. Vertesi, M. E. Lynch, & S. Woolgar (Eds.), *Representation in Scientific Practice Revisited* (pp. 201–221). Cambridge, MA: MIT Press.

Collins, H., & Pinch. T. (1998). *The Golem at Large: What You Should Know About Technology*. Cambridge: Cambridge University Press.

Forss, A. (2012). Cells and the (imaginary) patient: The multistable practitioner-technology-cell interface in the cytology laboratory. *Medicine, Health Care, & Philosophy, 15,* 295–308.

Friis, J. K. B. O. (2012). Perception: Embodiment and beyond. *Foundations of Science, 17,* 363–367.

———. (2015). Towards a hermeneutics of unveiling. In R. Rosenberger & P.-P. Verbeek (Eds.), *Postphenomenological Investigations: Essays on Human-Technology Relations* (pp. 215–225). Lanham, MD: Lexington Books.

Friis, J. K. B. O., & Crease, R. P. (Eds.). (2015). *Technoscience & Postphenomenology: The Manhattan Papers.* Lanham, MD: Lexington Books/Rowman Littlefield Press.

Haraway, D. (2003). *The Companion Species Manifesto*. Cambridge: Prickly Paradigm Press.

Hasse, C. (2008). Postphenomenology: Learning cultural perception in science. *Human Studies, 31*(1), 43–61.

———. (2015). *An Anthropology of Learning: On Nested Frictions in Cultural Ecologies*. Dordrecht: Springer.

Heidegger, M. (1996). *Being and Time*. (J. Stambaugh, Trans.). Albany, NY: SUNY Press. (Originally published 1927)

Howell, O. (2005). The "Creative Class" and the gentrifying city: Skateboarding in Philadelphia's Love Park. *Journal of Architectural Education*, *59*(2), 32–42.

Ihde, D. (1990). *Technology and the Lifeworld.* Bloomington, IN: Indiana University Press.

———. (1998). *Expanding Hermeneutics: Visualism in Science*. Evanston, IL: Northwestern University Press.

———. (2016). *Husserl's Missing Technologies*. New York, NY: Fordham.

Latour, B. (1992). Where are the missing masses? The sociology of a few mundane artifacts. In W. E. Bijker & J. Law (Eds.), *Shaping Technology/Building Society* (pp. 226–258). London: MIT Press.

———. (1999). *Pandora's Hope: Essays on the Reality of Science Studies*. Cambridge, MA: Harvard University Press.

———. (2005). *Reassembling the Social: An Introduction to Actor-Network-Theory*. Oxford: Oxford University Press.

Law, J., & Hassard, J. (1999). *Actor-Network Theory and After*. Oxford: Blackwell.

Lee, N., & Brown, S. (1994). Otherness and the Actor Network: The undiscovered continent. *American Behavioral Scientist, 37*(6), 772–790.

McQuade, D. (2016a, February 12). A farewell: LOVE Park, skateboard mecca. *Philadelphia Magazine*. 2.12.2016. http://www.phillymag.com/news/2016/02/12/farewell-love-park-skateboard-mecca/

———. (2016b, February 14). WATCH: That time Ed Bacon skateboarded across LOVE Park. *Philadelphia Magazine*. http://www.phillymag.com/news/2016/02/14/love-park-ed-bacon-skateboarding-video/

Merleau-Ponty, M. (1962). *Phenomenology of Perception*. (C. Smith, Trans.). London: Routledge & Kegan Paul. (Originally published 1945)

Mol, A. (2002). *The Body Multiple: Ontology in Medical Practice*. Durham, NC: Duke University Press.

Németh, J. (2006). Conflict, exclusion, relocation: Skateboarding and public space. *Journal of Urban Design, 11*(3), 297–318.

Oudshoorn, N. (2003). *The Male Pill: A Biography of a Technology in the Making*. Durham, NC: Duke University Press.

Oudshoorn, N., & Pinch, T. (Eds.). (2003). *How Users Matter: The Co-construction of Users and Technologies*. Cambridge, MA: MIT Press.

Pinch, T. J., & Bijker, W. E. (1984). The social construction of facts and artifacts: Or how the sociology of science and the sociology of technology might benefit each other. *Social Studies of Science, 14*(3), 399–441.

Pinch, T. J., & Trocco, F. (2002). *Analog Days: The Invention and Impact of the Moog Synthesizer*. Cambridge, MA: Harvard University Press.

Rosenberger, R. (2011). A case study in the applied philosophy of imaging: The synaptic vesicle debate. *Science, Technology & Human Values, 36*(6), 6–32.

———. (2013). Mediating Mars: Perceptual experience and scientific imaging technologies. *Foundations of Science*, *18,* 75–91.

———. (2014). Multistability and the agency of mundane artifacts: From speed bumps to subway benches. *Human Studies*, *37,* 369–392.

———. (2017a). Callous Objects: Designs Against the Homeless. Minneapolis: Minnesota University Press.

———. (2017b). On the hermeneutics of everyday things: Or, the philosophy of fire hydrants. *AI & Society*, *32*, 233–241. DOI 10.1007/s00146-016-0674-3.

Rosenberger, R., & Verbeek, P.-P. (Eds.). (2015). *Postphenomenological Investigations: Essays on Human-Technology Relations*. Lanham, MD: Lexington Books/Rowman Littlefield Press.

Saffron, I. (2002, August 23). Disappointing developments on 15th St. *Philadelphia Inquirer*, City-D Edition, p. D01.

Smith, A. (2003). Do you believe in ethics? Latour and Ihde in the trenches of the science wars (Or: Watch out Latour, Ihde's got a gun). In D. Ihde & E. Selinger (Eds.), *Chasing Technoscience* (pp. 182–194). Bloomington, IN: Indiana University Press.

Valenzuela, R. (2002, May 2–8). Bye bye Love. *City Paper*. http://archives.citypaper.net/articles/2002-05-02/naked.shtml

Vallor, S. (2015). Beyond ordinary givenness? Postphenomenology, digital imaging, and evidentiary responsibility. In J. K. B. O. Friis & R. P. Crease (Eds.), *Technoscience and Postphenomenology: The Manhattan Papers* (pp. 19–37). Lanham, MD: Lexington Books/Rowman Littlefield Press.

Verbeek, P.-P. (2005). *What Things Do*. University Park, PA: Penn State University Press.

———. (2011). *Moralizing Technology*. Chicago, IL: Chicago University Press.

Wiltse, H. (2014). Unpacking digital material mediation. *Techné, 18*(3), 154–182.

Chapter 10

Describing and Valuing Technological Mediation

From Postphenomenological Bridgeheads to Technoethical Outposts

Michel Puech

INTRODUCTION

Our current lifeworld can be considered philosophically uncharted territory. It is a technosphere: it is qualitatively and not only quantitatively different from the technological environments of human existence since Neolithic times. Here, I will not directly defend its disruptive character. I simply begin from the intuition of a specific technosphere representing a kind of intellectual frontier for philosophy. New territories are conquered (by the industry) with a fierce determination that may be resented on the "user side" as a violent intrusion into personal and social existence. This hectic frontier of our global civilization challenges the existing methods in human and social sciences. The societal demand for a pluralistic but definite set of norms requires methodologies that the academy is still struggling to delineate and to apply non-trivially to substantive issues. To focus on methods: we are good at describing and denouncing, but not so good at valuing and providing constructive guidance, particularly in ethics. How good we are at describing and analyzing technological uses is particularly substantiated by postphenomenological achievements in the last decades. The method for extending these achievements toward ethical assessment is already active in postphenomenology (Puech, 2016, sec. 1.1; Verbeek, 2011). I hope to contribute to this extension by using resources from recent virtue ethics and from Asian wisdom doctrines.

Postphenomenology offers an original method for assessing our elusive mediation with the technosphere, probably because it stands neither on the objective side (easily leading to technological determinism) nor on the subjective side (easily leading to "humanist" conservatism). Postphenomenology concentrates on the emergence of both "human" and "world" in their specific current avatars, a disoriented but frenetic Technosapiens and a tremendously vibrant but terribly ambivalent technosphere. From the vast range of subjects in this field of research, I address one precise issue: the *ethical* mediation in *ordinary* techno-human relationships. On this issue, postphenomenology can be complemented by other philosophical methods in order to expand its descriptive and pre-normative assessment capacity into a fully evaluative and hopefully constructive technoethics.

I will concentrate on one case study for making my points in this chapter: data immersion, that is to say the infosphere as a dense and active interface between subject and world. In this case study, one can easily imagine how standing on the objective side of the interface leads to functionalist (not to say business-oriented) and optimistic views, and how standing on the subjective side of the interface induces resistive (not to say politically prejudiced) and pessimistic views. Considering the interface as a generative mediation process, a postphenomological method in ethics has a chance to offer a constructive and specific view of Technosapiens's opportunities in the radically new infosphere.

THE EMPIRICAL AND PHENOMENOLOGICAL TURN: DESCRIBING AND UNDERSTANDING TECHNOLOGICAL MEDIATION

The main philosophical force in postphenomenology comes directly from classical phenomenology, the apparently simple capacity to describe. Descriptive assessment is a precious method, sometimes skipped in the rush to "facts," numbers, statistics, and all traditional forms of empirical evidence. Radical description is made possible by the method of *epochē*, the radical suspension of judgment. The most pressing issues in our civilization of abundance (material and informational) are linked to abundance itself, more precisely to its incessant material and informational solicitations. Pre-existing interpretative systems (cognitive, symbolic, political, ethical) are always already active and prejudicial in the "natural" attitude. How can we maintain this methodological descriptive strength and put it at the service of evaluation, norms, and the enlightenment of practical decision-making? Does radical description imply a definitive ban on value assessment? Or can it be a methodological step that allows a stronger evaluation and even a form of ethical commitment? I opt for the latter scenario.

Merleau-Ponty's introduction to the phenomenological method has been influential for postphenomenology. To what he dubbed "traditional prejudices," he directly opposed the radical description motto of phenomenology, "return to phenomena" (Merleau-Ponty, 1945, Introduction). Are these prejudices cognitive or ethical? Far more cognitive than ethical according to mainstream phenomenology. Thus, the phenomenological method, from Husserl on, is predominantly an epistemic reform. But even in Merleau-Ponty's phenomenology of perception, not to mention in Heideggerian phenomenology, cognitive or ontological prejudices are linked to evaluation issues. These were not considered relevant to the core of the phenomenological method in its first form, in either the German or the French schools.

In postphenomenology, which is essentially a philosophy of technology, the subject of "technology" introduces a more interesting connection: the cognitive and ontological prejudices concerning technology always carry the threat of a prejudicial evaluation of technology. Resisting this threat is a challenge. This axiological issue at the heart of the phenomenological method is clearly perceptible in recent positions on postphenomenological methods (Thomson, 2009; Mitcham & Waelbers, 2009; Rosenberger & Verbeek, 2015, pp. 30–32).

The methodological advantage of postphenomenology consequently lies in the combination of a paradigm-neutral descriptive capacity with a specific sensibility to questions of value. There is no paradox here because these value questions emerge as questions only when the dominant interpretative systems are turned off. To turn them off remains the unique strength of a phenomenological style in philosophy, provided it sticks to ordinary empirical uses. The ethical strength of posts-phenomenological methods is the suspension (*epochē*) of the overwhelming hermeneutic models, particularly the two most pregnant ones: the intellectual routines of "social critique" of the 1970s and at the other extreme the gullible eagerness for consumption. How can we see the new smartphone as neither a manipulative capitalist abomination nor a super-powerful innocent toy? How can we stop seeing it as both, actually? And why should we? How can we just *see* it, to start with?

In terms of "turns," a term readily used by contemporary trends of philosophy to insist on their own originality, postphenomenology celebrates the "empirical turn" (Achterhuis, 2001; Kroes & Meijers, 2001). At first sight this turn implies more mundane description and less critical assessment, an alternative corresponding for philosophers to the famous Humean gap between the descriptive "is" and the normative "ought." In reality, the empirical case studies in recent literature criticize the pseudo-neutrality and the pseudo-autonomy of technology. Against the previously essentialist and deterministic approaches to technology, the "empirical turn" sets a new stage where the "naked" description of technological mediation is paramount and it becomes immediately instructive for philosophers. Further, there is a continuity between this empirical turn in the 1990s and an "ethical turn" in the 2000s (Puech, 2016, sec. 1.1; Verbeek, 2011, p. 160). Richard Rorty is a reference on the pragmatic transition from ontological critique to ethical views. In Ihde's dialogue with Rorty, tellingly, the non-essentialist and non-paternalistic orientation of postphenomenological methods is at the forefront (Ihde, 1986, Introduction, Chapter 9). A third turn is then conceivable and is equally important for the methodology of technoethics: the pragmatic turn. It has received its share of attention (Keulartz, Korthals Schermer, & Swierstra, 2002) and its symbiosis with postphenomenology is influential in recent technoethics.

All these methodological considerations are not only of historical and epistemological importance. The objective is to pin down the technosphere in its ontological, existential, and ethical specificity. There is a "disseminated frontier" of our global civilization; it is everywhere, ubiquitous and incessantly reshaping human existence and communities. This frontier lies in the *mediation* with the technosphere. In this mediation process Technosapiens and the technosphere conjointly emerge. Postphenomenology's basic framework, human/world/ technology pattern, is exactly situated for apprehending this process.

Between Husserl and Borgmann, there is a clear continuity in understanding the human/world emergence through the concept of lifeworld (*Lebenswelt* in German). Husserl's focusing on the *Lebenswelt* was a complex moment in his reform and refoundation of logic and psychology. One of its most remarkable progeny is the young Heidegger's take on phenomenology, centered on "life" and "lifeworld" in his 1919/1920 lectures on phenomenology (Heidegger, 1993). Before he took Being as his central theme, Heidegger meditated on the methods for a radical description of the lifeworld, elaborating the concepts of a new existential analysis—for instance "Weltcharakter des Lebens" (world-character of life) (Heidegger, 1993, p. 33) or "Grunderfahrung der Selbstwelt" (fundamental experience of the self-world) (Heidegger, 1993, p. 93). Borgmann's path-breaking work on the technological lifeworld (Borgmann, 1984) is on this trajectory. This whole method can be recaptured today as a method of *mediation* analysis and applied to the new mediation between human and the contemporary technosphere and infosphere.

Ihde's isolation of the technological lifeworld as a philosophical subject of its own right (Ihde, 1979) inaugurated the postphenomenological "mainstream" descriptive methods with an open-minded and pluralistic understanding that clears the way for bottom-up pragmatic evaluations. This evaluation remains implicit in most of Ihde's works. However, in these descriptive methods there is a nascent technoethics that will evolve into a (falsely) candid philosophical "accompaniment" of technology (Puech, 2016, pp. 21–22; Verbeek, 2011, 2013). Ihde initially issued a warning: "However, both the ethical and the social-political questions, as urgent and as important as they are, deal with *effects* of technology rather than examine the phenomenon itself" (Ihde, 1979, p. xxiii). "Back to the things themselves" means then in a first step: receding from evaluation and examining the phenomenon itself. Ironically, this moment leads not to neutrality but to an idea of non-neutrality: "any use of technology is non-neutral, it transforms experience" (Ihde, 1979, p. 53). Therefrom comes a typical hermeneutical non-neutrality, very Merleau-Pontyan and epistemic in its descriptive beginnings, but with a potential for a renewed evaluation of what it describes: the non-neutral transformation of human experience. Ihde's views can thus be rendered in a characteristically pre-ethical perspective (Verbeek, 2011, p. 14ff.). Avoiding both essentialist and instrumentalist approaches, postphenomenology came to determine *mediation* as the key concept for understanding the technosphere (Verbeek, 2005, p. 11). There is a specific intentionality in technology mediation (Verbeek, 2011, p. 57).

Concerning *data immersion*, the unapparent reality of hardware, software, and code behind our superficial experience of the infosphere maintains a typical opacity in digital mediation—for the simple reason that most people have absolutely no idea of what code is and does. Therefore, it is not surprising to find *phenomenological* axioms and methods in a critical book on data and code: Berry introduces his "philosophy of code and mediation" as "a phenomenological approach that tries to highlight the pragmata of code" (Berry, 2011, p. 5). He even suggests "a phenomenology of computation" (Berry, 2011, Chapter 5). Code is the unapparent entity that shapes our mediation with the infosphere and then with the technosphere. What we have to understand in our ordinary interfacing with the infosphere is so close to us and so pervasive that it requires a specific methodological effort to reach it—"I want to suggest that what is happening in the 'digital age' is that we increasingly find a computational dimension inserted into the 'given.' Or better, that the ontology of the computational is increasingly hegemonic in forming the background presupposition for our understanding the world" (Berry, 2011, p. 128). There are unapparent intentionalities, human and nonhuman, in digital mediation. But here the intentionality patterns are far from the straight line of the perceiving eye and its perceptible object (the classical paradigm). Intentionalities and hermeneutic correlations in the digital world are even more intricate, hyper-complex, disseminated, "ambient" (as in "ambient intelligence") than in the original instrumental relationships described in postphenomenology. Data immersion is not the sum of additive instrumental mediations, it plunges the digital Dasein into a sort of ambient sphere where mediation is the substance of technology. The new features in this ambient mediation are data persistence and "searchability," ubiquitous services, profiling, and of course a boundless surveillance potential.

ORDINARY TECHNOLOGICAL MEDIATION AS AN ETHICAL ISSUE

STS (science, technology, societies) studies, in their sociological and quasi-positivist variants, have a tradition of resistance and even open hostility to questioning values, norms, ethics (Keulartz, Korthals Schermer, & Swierstra, 2002, Introduction, Chapter 1). Postphenomenology has found for itself a way from the descriptive to the normative, particularly in examining the *ordinary* in our mediation with the technosphere. As there is always already meaning and intentionality in our mediation with the world there is always a potential value-attitude in all these ordinary intentionalities. Bridging the descriptive/normative gap has deep motivation in the ancestry of postphenomenology.

In Spiegelberger's classical survey of the original phenomenological movement (Spiegelberg, 1965) several places can be discerned where ethical assessment emerges even though it eventually reintegrates the flow of mainstream phenomenology most of the time, remaining focused on ontology and epistemology. Brentano's psychological deduction of self-evident and objective moral principles, for instance (Spiegelberg, 1965, pp. 46–48), remains in a cognitivist style and far from applied ethics and pragmatics. Max Scheler unquestionably engages an ambitious phenomenology of ethics and value, but the really ethical content remains in the form of a cognitivist analysis (Spiegelberg, 1965, p. 254). The same option commands his influential phenomenology of sympathy, a moral emotion. However, ethical issues become more and more important in the history of the movement as it drifts away from Husserl: Nicolai Hartmann's *Ethik* introduces a phenomena of value "discovery" (Spiegelberg, 1965, p. 384), and French phenomenology, with the "existentialist" movement, was deeply committed to values—moral, social, and political values. This is the case for Merleau-Ponty as much as it is for Sartre.

In this historical evolution a growing consideration for the ordinary in human experience can be discerned. The transition from description to evaluation, where postphenomenology has something important to contribute, does not travel the high road of "essences" but rather the winding path of existences. In Ihde's original research about vision and audition, the two dimensions (ontological-epistemic and ordinary lifeworld) are in symbiosis, sometimes seamlessly incorporated into the progressing analysis, sometimes investigated in Merleau-Pontyan style. This is not the right place for tracking and commenting such passages in Ihde, but they justify this important methodological point made by Mitcham and Waelbers: "The theories of ethical drama that have been associated here with Winner, Latour and their followers are cumulative products of a trajectory of dialogue on relations between technology and ethics that remain rich but largely descriptive in character" (2009, p. 379). Ihde himself tackles the descriptive/normative issue in his self-commentary (2006, p. 277). There is more in Mitcham's and Waelbers's appreciation: "Descriptive ethics is a necessary prolegomenon to prescriptive ethics" (2009, p. 379). The notion of "descriptive ethics" offers in itself a simple conceptual bridge between the descriptive and the normative. Descriptive ethics scrutinizes the ordinary mediation with technology, on the existential level, which stands in the intermediary position between hermeneutic concepts and vocabulary, on the one side, and explicitly ethical considerations, on the other side. The transition table hermeneutic/existential in Verbeek (2005, p. 196) can be extended with an ethical column.

Verbeek's *Moralizing technology* (2011) substantiates the debate on the descriptive/normative issue in philosophy of technology. His notion of "mediated morality" reinvests STS methods and concepts (which are essentially Latourian) by examining a specific intentionality in technology mediation (Verbeek, 2011, p. 57). Beyond the classical STS loci on designing and using artifacts, this approach is fertile for delineating a specific ethics of *ordinary* mediation with technology. The descriptive/normative gap is uniquely bridged by considering the ordinary: this is my thesis and I contend that it can be perceived in the evolution of postphenomenology itself.

The "material turn" that stimulates ethics according to Verbeek (2006, p. 117) is a move toward the ordinary, not only the ordinary of humans (as opposed to gods or essences) but the ordinary of *things*. An ethics of mediating with things was strangely lacking in Western philosophy. Its emergence is linked to the theoretical importance of the ordinary in several fields of science and culture: Quine on language and logic (Quine, 1960), Varela on the neurological basis for ordinary moral skills (Varela, 1999), Illich on the ethical and political importance of the "vernacular" (Illich, 1980), Dreyfus on "the phenomenology of everyday expertise" and the "implications of the phenomenology of expertise for ethical experience" (Dreyfus, 2014) (see more about the ordinary in contemporary technoethics in Puech, 2013; Puech, 2016, Chapter 3).

Addressing the ordinary is key for a phenomenological passage between description and ethical assessment because the very nature of the ordinary is to be unseen: unseen by philosophers for sure but also unseen by the ordinary ethical agent himself/herself. Digital modernity means dealing with unseen mediations which are disseminated and always active everywhere in the existential sphere of individuals and in social common space. Phenomenology is primarily concerned with phenomena that remain "hidden in plain sight" and contemporary technology is full of them, says Ian Thomson (2009, p. 195ff.). Technoethics is often confronted with a situation where nothing can be assessed simply because nothing is adequately described. This analysis applies to technological mediation at every stage and it applies particularly well to data immersion. On a well-known level it illuminates the universal use of artifacts without any trace of the corporate and marketing "scripts" intentionally designed into them. This point leads to a critique of the technological society in STS studies. But the descriptive and then normative consideration of phenomena "hidden in plain sight" can also lead to a constructive ethics of individual and collective agency. Concerning data immersion, the unseen and "scripted" pervasive presence of data does not imply their essential malfeasance, as it is assumed when no "descriptive ethics" of digital mediation is seriously attempted.

More than yet another critique of the technosystem, Borgmann's book on the "character of contemporary life" *assesses* technology and particularly ordinary technology, the latter being of greater existential pregnancy and importance in his approach, even if this orientation is not explicitly thematized. Borgmann crosses the descriptive/normative separation to reach "ordinary life." His plea for this transition is outspoken in his defense of "deictic discourse" (Borgmann, 1984, Chapter 21), resisting "the immunity of technology to traditional moral analysis" (Borgmann, 1984, p. 169). In Borgmann "deictic" means a "discourse of ultimate concern," "drawing continued strength from something that is present visibly, forcefully, and in its own right," and "addressing others by inviting them to see for themselves" (Borgmann, 1984, p. 178). Deictic discourse must be "restored," for the sake of democracy in particular (Borgmann, 1984, p. 177). Borgmann's contribution to this restoration resides largely in his theory of *focal* things and practices, which can be read as a masterful phenomenology of ordinary existence in technology. "Focal" (from *focus*, meaning "hearth") in Borgmann is the opposite of commodities and simple devices, it points to the objects and moments that can be "centers" of meaning, concern, intrinsic value in a human life. The symbiosis between descriptive and normative methods in Borgmann's work justifies Mitcham and Waelbers's conclusion, affirming that Borgmann "[provides] a paradigm of descriptive sensitivity woven together with an enriched and enriching normative seriousness" (2009, p. 381).

In the evolution from Ihde's dental probe to Verbeek's obstetric ultrasound to Rosenberger's smartphone using while driving the car (Rosenberger, 2014), more than just the normative/descriptive gap is bridged. A transition is perceptible, from simply assessing to positively valuing some options in our technological behaviors. Philosophers must be in a position to address the confused feeling of a *moral* value linked to open source software, for instance. This transition leads to a constructive technoethics. Its most innovative feature, the project of "accompanying" technology, is a direct effect of the emergence of the ordinary as a key dimension in assessing technological mediation. My recurrent case study helps to make this point.

In data immersion the invisible and pervasive character of ordinary mediation reaches its climax. Leaving "traces" everywhere on the Web is not without moral importance, but we need an inquisitive and obstinate descriptive phenomenology of ordinary digital mediation for gaining the capacity to *morally* assess our footprint in the infosphere. Here again, attention is focused on some all too visible corporations and their technological devices while the micro-actions and micro-resistance of ordinary users make the difference, or can make a difference. These micro-actions are specific mediations, blended and mingled into the ordinary infosphere. Every click on a link has moral significance for this is the very moment of mediation with the corporations and administrations that are out of reach in the contemporary world—out of material reach but more importantly out of political reach. Delineating the multiple intentionalities in the networks of humans and nonhumans that are the current infosphere means for instance following the existential investment in Facebook or in Pokémon Go chasing. This existential investment is a mixed mediation between "real" world and digital data. It is more than multistable in the accepted postphenomenological sense of the word, it is multi-focal in a Borgmannian sense, and more than anything it engages a pluralist ethic, ambiguous as Boyd has shown (2015), multilayered and versatile. Exploring the values system of this kind of mediation is different from any functional analysis, be it technological or behavioral, it calls for an exploration of the existential ordinary that still requires philosophy, in a postphenomenological style.

VIRTUE ETHICS AND TECHNOLOGIES OF THE SELF

The renewed methods that illuminate ordinary technological mediation builds technoethical outposts in philosophically uncharted territory. Locating some of these outposts can help them to mutually support themselves and to thereby reinforce our intellectual tools for morally assessing the technosphere. A sequence of three stages can be useful for expanding postphenomenology toward technoethics: from describing to assessing, from assessing to valuing—in a positive and constructive meaning of this latter term. Envisioning this last stage now requires reconsidering some methods in the human and social sciences.

When trying to assess electricity consumption, an article by Geerts (2012) raises a relevant question of "self-practice." The transparency that prevents one from assessing energy consumption lies on the side of the subject, the human person, the self who uses devices in the most ordinary context. In the end, and very phenomenologically, what is transparent in the ordinary even more than ordinary life itself is the self, the human person. The eclipse of the self in philosophy (due to a complex evolution in 20th-century philosophy that I will naturally neither explain nor document here, but see Puech, 2016, Chapter 4) deprives the assessment of technology of one ethically significant option: valuing technologies in their self-building potential. As long as the self is interpreted as a resulting equilibrium of forces of social determination, the outcomes of assessing technology fuels social critique without limitation. A constructive valuing of technology makes sense when the ordinary self is rehabilitated with a genuine capacity for authenticity, self-determination, and all the "humanist" values.

Postphenomenology has a role in this reorientation of philosophy. Certain methodological aspects of postphenomenology can lead to the notion of an ethically self-constructive mediation with technology, in recovering a philosophically acceptable notion of the self. The first one is the insistence on "situated" analysis in phenomenology and postphenomenology. One of these situations, ordinary life, reveals some aspects of technological mediation that are not incidental. Another methodological breakthrough is the "non-foundational" approach, typical of postphenomenology (Ihde, 1993, p. 1). This is a break with Husserl and the obsession with foundation in Continental philosophy since Descartes. The situation is not so simple, however, because the phenomenological "ego" remains under the name "self" in the mediation scheme *human/world/technology*. That which in the postphenomenological framework is described as a web of intentionalities that co-constitute human, world, and technology is more and more often approached in terms of "technologies of the self" under the influence of Foucault.

Foucault's last publications and lectures introduced a philosophy of the subject which is a philosophy of mediation. Foucault tries to show how the self constitutes itself in the constant process of being shaped by domination and of resisting this domination on the very concrete level of micro-behaviors and micro-experiences. The method reinvented by Foucault in this attempt explicitly draws from ancient philosophy, Stoicism in particular, but it also inherits from the phenomenological spirit prevailing in the Parisian intellectual circles where Foucault was initially trained. For these reasons, Foucault's concept of the self in his last lessons is no longer the politically dominated "subject" of Foucault's earlier works, and a postphenomenological interpretation of this self is not a blunt misinterpretation. Consequently, an ethics of technological mediation with Foucauldian sources thrives in postphenomenological philosophy of technology (Dorrestijn, 2012; Verbeek, 2011, Chapter 4). Technological mediation is methodologically situated in the broader context of the subjectivation process in Foucault as in postphenomenology. Foucault's initial "techniques of the self" actually involved some technology and artifacts, a pen and a notebook for writing a private journal for instance: in the digital technosphere of today the booming devices for the "quantified self" and all sorts of sophisticated technological assistance have transformed this strange idea of "technologies of the self" into one of the most pervasive and controversial lifestyle. Foucault's style of existential analysis in reading the Hellenistic schools of moral philosophy is a paradigm for reading the ordinary life in the present technosphere. It refers to original Greek words (*technē, epimeleia, parrhesia*) that are currently being added to the list of Latin, German, and improbable neologisms populating philosophical parlance.

Inspired by Foucault, De Vries examines the ethical consequences of a phenomenological-pragmatic analysis of ordinary medical practice in terms of "truth games" and "power relations": the descriptive method here is key for a specific approach to normativity and for the critical assessment of non-visible norms in medical practice (De Vries, 2002). Health mediations happen in the technosphere and infosphere now. There is a way of describing visible mediation that reveals its invisible norms. In a first step, this method is a powerful tool for social critique and the deconstruction of inauthentic subjectivity. But in a second step a constructive valuation of ordinary technological mediation could make sense. Philosophers have to face the ambiguous potential of ordinary technologies: the same device that tracks my data and sells them to corporations gives me free access to virtually the whole of culture—and incidentally leads me safely back to my hotel in an unknown place. Doesn't this ordinary virtuosity deserve some valuation in spite of the background ambiguity built-in to the device? Descriptive ethics here must precede normative ethics and postphenomenological restraint helps to contain prejudiced conclusions.

A slightly embarrassing notion surfaces in some of these researches: *virtue*. The ethics of virtue is one of the most productive trend in current moral philosophy and it could be the main opportunity for shifting to constructive ethics. But in our secular culture the term "virtue" is still suspect of shallow and preachy moralism. Surmounting this pejorative connotation, virtue ethics has already been able to connect with different sets of useful concepts in at least two cases: connecting with the postphenomenological/Foucauldian approaches, on the one hand, and on the other hand connecting with certain Asian worldviews that offer a global alternative to the Western industrial worldview.

In a chapter trying to bridge "Phenomenology and Virtue Ethics," Reynolds "argue[s] that there is both a methodological proximity and, to a lesser extent, a substantive ethical proximity between phenomenology and virtue ethics around the centrality of character to moral flourishing" (Reynolds,

2013, n.p.). Indeed, ordinary technologies of the self require exactly the kind of "reflective view" that defines the specific *perspective* of virtue ethics and its practical wisdom (Tiberius, 2008). In this line of thought, the self is not guided by deduction from higher principles, but on the contrary it is seized in the mundane details of everyday existence: a situated moral self is examined as a self-constituting entity, from a situated point of view, from inside a practice. I need not stress the methodological isomorphism with postphenomenology. On this matter mediation with the technosphere counts as one of the most characteristic practices for the self-constitution of contemporary human persons. The transparency of the intentionalities involved in this constitution process is at the same time hermeneutic transparency (what are the implied meanings?) and ethical transparency (what are the implied values?). This transparency is in fact a philosophical opacity that postphenomenological methods intend to dispel, descriptively and then ethically. The alliance with virtue ethics' own methods and concepts is key. For instance, Slote's (2007) analysis of empathy is a masterpiece in recent virtue ethics, interpreting *care* as the universal vector for morality and values. But this care typically concerns humans only, not objects and artifacts. What we feel for our car or computer is obviously a morally significant element in our existence but most of the time it remains invisible to philosophers as long as no other human is directly involved. What can be accomplished when the scope of ethico-phenomenological attention is enlarged to the ordinary technosphere can be glimpsed through the help of Slote's ideas on virtues like moderation and satisficing (Slote, 1989, Chapter 1). Promising case studies concerning "moderation" in car driving or a "satisficing" use of Amazon (a quintessentially digital case study) can take inspiration from Slote's situated cases—second cups of tea, so English, or snacking all day long, so American (Slote, 1989, pp. 12–13).

Technologies of the self and virtue ethics merge within a totally new methodological context when Asian studies enter the debate. The philosopher of reference on this matter is Peter D. Hershock. His research on applying Buddhist concepts to the moral assessment of the technosphere (Hershock, 1999) parallels postphenomenology in concentrating on the details of practical life and investing them with a crucial moral significance. When Ihde stated that "What is needed is what I shall call a 'loose' or maybe even 'Zen' relation to technology" (Ihde, 1983, p. 23), he certainly had in mind Heidegger's *Gelassenheit* (serenity as a free relationship to technology) but perhaps not the actual Zen practice of *samu* (作務) in Japanese culture: meditative domestic work (Ives, 1992; Puech, 2016, pp. 165–168). Behind these Japanese and Zen versions which fit particularly well the ordinary technosphere, classical Chinese philosophy demonstrates the possibility and the value of a meditative attitude for ordinary life in its material dimension (Fingarette, 1972).

Applying these methods and notions to *data immersion* shows a lack of balance in the previous approach: the dangers for the self have been abundantly reported but valuing the constructive opportunities for the self of quasi-universal data access remains to be philosophically defended. This lack of balance is largely due to the eclipse of the self and of self-care in philosophical considerations. In a data-rich technosphere, the self needs opportunities for learning and practicing virtue no less than in any other environment, with this unexpected difficulty: this environment offers no points of reference, no stable value, for making the best out of digital opportunities. Deploying "digital literacy" remains largely an instrumental approach and what is called "digital humanities" for now thrives essentially on paper, in applications for research grants, not in common culture. Turning digital technology into techniques of a flourishing self is possible by reorienting the methods of philosophy of technology, particularly the postphenomenological expertise for describing the ordinary, toward a clearly endorsed technoethical project. This requires interdisciplinary collaboration in philosophy at a really large scale. Here postphenomenology is at a crossroads, in my humble opinion (and practice): seizing the opportunity to hybridize itself into a pluralistic method or else maintaining a strong methodological orthodoxy. This chapter and this section show well enough where I stand concerning this choice.

TECHNOETHICS OF ORDINARY MEDIATION: THREE PRINCIPLES TO START WITH

A constructive technoethics is already taking shape inside postphenomenology or on its fringes. I will take as evidence for this point the "Constructive Technology Assessment" method proposed by Verbeek and his conclusion on "accompanying technology" in the specific perspective of the good life (Verbeek, 2011, p. 102, Chapter 8). My intention in this chapter was to outline and to justify a methodological inflection, namely toward the most *ordinary* personal mediation with the technosphere and a strictly *ethical* evaluation of it. Possible outcomes can be prospectively summarized by a set of three principles.

1. *Individual pragmatism*: In assessing ordinary technological mediation, this means sticking to the lower level of human interface with the technosphere and on this level practicing a strict *epochē* (suspension of judgment) on the always already pregnant devaluation of the ordinary. This descriptive method opens the way to *valuing* some usually neglected mediation with technology.
2. *Flourishing* as the key personal value. In assessing ordinary technological mediation, this means conferring on the individual self (and possibly other living entities or local/global ecosystems) an intrinsic value, making it a "focal" entity in a Borgmannian sense. In situated descriptions, this particular situation centered on a self-constituting entity changes the

reference system. More precisely, the reference system translates from institutional politics and economics into informal practical wisdom and micro-practices.

3. *Harmony* as the paramount collective value. In assessing ordinary technological mediation, this means reintegrating the global (politics, the economy, and all that is candidly "suspended" by the micro-phenomenological effort), but not as a determining system with domination/resistance relations between humans and technologies. Harmony rather insists on non-confrontational relations and non-deterministic balance. The biological image of an ecosystem comes to mind for illustrating this process through which value is created by the aggregation of multiple micro-flourishing. The Internet is another telling illustration, stressing the fact that harmony as a value does not mean perfection, as in "the best possible world" conceived by classical metaphysics. Neither Darwinian nature not the actual Internet are ontologically and ethically perfect worlds. But they are harmonic worlds. Describing harmony and particularly describing how micro-actions can harmonize and propagate harmony leads to a differentiated evaluation of the technosphere and infosphere—using for instance robust moral notions in favor of open source software, self-sustaining ethical notions that are not the consequences of preexisting politico-economic agendas.

Principles (1) and (2) are linked to recent virtue ethics and wisdom studies (broadly construed), principles (2) and (3) are linked to revisited Asian traditions of thought. When examining the methodological needs for descriptive and evaluative accounts of contemporary technology, Val Dusek affirms:

> By forcing the integration of ethics and political philosophy with epistemology and philosophy of science, as well as inviting the mutual employment and combination of the methods of logical and linguistic analysis with phenomenology, hermeneutics, social constructionism and process philosophy, philosophy of technology will move from being a marginal and neophyte specialty to playing a central role in the cross-pollination of both subject fields and methods of contemporary philosophy leading to a reunited world philosophy community. (2009, p. 134)

While saluting this utopian vision, technoethics, a branch of philosophy of technology, is an even smaller actor, but its privileged perspective on the technosphere makes it a relevant actor for combining philosophical methods and inspirations. Pragmatically, it appears that the contemporary Technosapiens has a deficient perception of the artifacts that operate extremely near to him/her: one's mobile phone, the computer screen in front of which one spends his/her working and leisure time, and more globally the technosphere that is nearer to us than anything traditionally considered by philosophers. Convergent philosophical methods facilitate the delicate transition from *describing* (in a phenomenological acceptation) this novelty to *valuing* it (aiming at a constructive applied ethics).

REFERENCES

Achterhuis, H. (2001). *American Philosophy of Technology: The Empirical Turn*. Bloomington, IN: Indiana University Press.

Berry, D. M. (2011). *The Philosophy of Software: Code and Mediation in the Digital Age*. Basingstoke & New York, NY: Palgrave Macmillan.

Borgmann, A. (1984). *Technology and the Character of Contemporary Life: A Philosophical Inquiry*. Chicago, IL: University of Chicago Press.

Boyd, D. (2015). *It's Complicated: The Social Lives of Networked Teens*. Yale University Press. Retrieved from http://www.danah.org/books/ItsComplicated.pdf

De Vries, G. (2002). Pragmatism for medical ethics. In F. W. J. Keulartz, M. Korthals, M. Schermer, & T. E. Swierstra (Eds.), *Pragmatist Ethics for a Technological Culture* (pp. 151–164). Dordrecht: Springer.

Dorrestijn, S. (2012). *The Design of our Own Lives Technical Mediation and Subjectivation After Foucault*. Enschede: University of Twente. Retrieved from http://www.stevendorrestijn.nl/downloads/Dorrestijn_Design_our_own_lives.pdf

Dreyfus, H. L. (2014). *Skillful Coping: Essays on the Phenomenology of Everyday Perception and Action* (M. A. Wrathall, Ed.). Oxford: Oxford University Press.

Dusek, V. (2009). Introduction: Philosophy and technology. In J. K. B. Olsen, S. A., Pedersen, & V. F. Hendricks (Eds.), *A Companion to the Philosophy of Technology* (pp. 131–140). Chichester & Malden, MA: Wiley-Blackwell.

Fingarette, H. (1972). *Confucius: The Secular as Sacred*. New York, NY: Harper & Row.

Geerts, R.-J. (2012). Self-practices and the experiential gap: An analysis of moral behavior around electricity consumption. *Techné: Research in Philosophy and Technology, 16*(2), 94–104.

Heidegger, M. (1993). *Grundprobleme der Phänomenologie (1919/20)—Gesamtausgabe Bd 58*. (H.-H. Gander, Ed.). Frankfurt am Main: V. Klostermann.

Hermberg, K., & Gyllenhammer, P. (Eds.). (2013). *Phenomenology and Virtue Ethics*. London & New York, NY: Bloomsbury.

Hershock, P. D. (1999). *Reinventing the Wheel: A Buddhist Response to the Information Age*. Albany, NY: State University of New York Press.

Ihde, D. (1979). *Technics and Praxis*. Dordrecht & Boston, MA: Reidel.

———. (1983). *Existential Technics*. Albany, NY: State University of New York Press.

———. (1986). *Consequences of Phenomenology*. Albany, NY: State University of New York Press.

———. (1993). *Postphenomenology: Essays in the Postmodern Context*. Evanston, IL: Northwestern University Press.

———. (2006). Forty years in the wilderness. In E. Selinger (Ed.), *Postphenomenology: A Critical Companion to Ihde* (pp. 267–290). Albany, NY: State University of New York Press.

Illich, I. (1980). Vernacular values. *CoEvolution Quarterly*. Retrieved from http://www.preservenet.com/theory/Illich/Vernacular.html

Ives, C. (1992). *Zen Awakening and Society*. Honolulu, HI: University of Hawaii Press.

Keulartz, F. W. J., Korthals, M., Schermer, M., & Swierstra, T. E. (Eds.). (2002). *Pragmatist Ethics for a Technological Culture*. Dordrecht: Springer.

Kroes, P., & Meijers, A. (2001). *The Empirical Turn in the Philosophy of Technology*. Amsterdam & New York, NY: JAI, Elsevier Science.

Merleau-Ponty, M. (1945). *Phénoménologie de la perception*. Paris: Gallimard.

Mitcham, C., & Waelbers, K. (2009). Technology and ethics: Overview. In J. K. B. Olsen, S. A., Pedersen, & V. F. Hendricks (Eds.), *A Companion to the Philosophy of Technology* (pp. 367–383). Chichester & Malden, MA: Wiley-Blackwell.

Olsen, J. K. B., Pedersen, S. A., & Hendricks, V. F. (Eds.). (2009). *A Companion to the Philosophy of Technology*. Chichester & Malden, MA: Wiley-Blackwell.

Puech, M. (2013). Ordinary technoethics. *International Journal of Technoethics*, *4*(2). https://doi.org/10.4018/jte.2013070103

———. (2016). *The Ethics of Ordinary Technology*. New York, NY: Routledge.

Quine, W. V. O. (1960). *Word and Object*. Cambridge, MA: Massachusetts Institute of Technology.

Reynolds, J. (2013). Phenomenology and virtue ethics: Complementary anti-theoretical methodological and ethical trajectories? In K. Hermberg & P. Gyllenhammer (Eds.), *Phenomenology and Virtue Ethics* (pp. 113–131). London & New York, NY: Bloomsbury.

Rosenberger, R. (2014). The phenomenological case for stricter regulation of cell phones and driving. *Techné: Research in Philosophy and Technology*, *18*(1/2), 20–47.

Rosenberger, R., & Verbeek, P.-P. (Eds.). (2015). *Postphenomenological Investigations: Essays on Human-Technology Relations*. Lanham, MD: Lexington Books/Rowman Littlefield Press.

Selinger, E. (Ed.). (2006). *Postphenomenology: A Critical Companion to Ihde*. Albany, NY: State University of New York Press.

Slote, M. (1989). *Beyond Optimizing: A Study of Rational Choice*. Cambridge, MA & London: Harvard University Press.

———. (2007). *The Ethics of Care and Empathy*. London & New York, NY: Routledge.

Spiegelberg, H. (1965). *The Phenomenological Movement: A Historical Introduction* (2nd ed.). The Hague: M. Nijhoff.

Thomson, I. (2009). Phenomenology and technology. In J. K. B. Olsen, S. A., Pedersen, & V. F. Hendricks (Eds.), *A Companion to the Philosophy of Technology* (pp. 195–201). Chichester & Malden, MA: Wiley-Blackwell.

Tiberius, V. (2008). *The Reflective Life: Living Wisely with Our Limits*. Oxford: Oxford University Press.

Varela, F. J. (1999). *Ethical Know-how: Action, Wisdom, and Cognition*. Stanford, CA: Stanford University Press.

Verbeek, P.-P. (2005). *What Things Do: Philosophical Reflections on Technology, Agency, and Design*. University Park, PA: Pennsylvania State University Press.

———. (2006). The morality of things: A postphenomenological inquiry. In E. Selinger (Ed.), *Postphenomenology: A Critical Companion to Ihde* (pp. 117–130). Albany, NY: State University of New York Press.

———. (2011). *Moralizing Technology: Understanding and Designing the Morality of Things*. Chicago, IL: University of Chicago Press.

———. (2013). Resistance is futile. *Techne: Research in Philosophy and Technology*, *17*(1), 72–92.

Chapter 11

Technological Mediation and Sociocultural Variability

Arun Kumar Tripathi

INTRODUCTION

Philosophy of technology promises the possibility of an understanding of technology that may be important not only to public policy but also in helping to conceptualize intellectual approaches to the study of technology and, indeed, to shaping new fields of knowledge and research. Philosophy of technology may also have a role to play in relation not only to structuring a largely disparate and inchoate field but also more directly in teaching and learning about technology (Peters, 2006, p. 96).

Technology is a social and political force. The technological devices and instruments such as telephones and specs we make and use, transform our experience in ways that are philosophically relevant. Technologies such as automobiles, telephones, and specs enlarge and extend our capacities and effects of changes in the natural and social worlds. Peter-Paul Verbeek (2001) argues, "Technology cannot be grasped in isolation; neither can culture" (p. 133). Technology always exists in its cultural contexts.

From a Euro-American perspective, technology is viewed through its connection with the sciences, while in South America the perspective is the reverse, science is viewed through its technologies understood as cultural instruments—this places the technification of sciences in the foreground. Don Ihde is one representative of phenomenology of technology that has demonstrated a willingness to connect North American and German traditions, specifically Martin Heidegger (who was clearly one of the forefathers of 20th-century philosophy of technology). Ihde understands technological development in terms of a social anthropology of technosystems. This viewpoint is in opposition to the technological determinism of applied natural science (Ihde, 1990, p. 5).

In contemporary continental philosophy of technology some questions are playing a pivotal role. We encounter questions such as: How is human behavior and embodiment affecting social and cultural factors? How do we relate to technologies in the lifeworld? What kind of relationships are there to technologies? And how does the lifeworld shape technologies and how does technology shape the lifeworld? Human experiences of our lifeworlds are shaped by physical and symbolic tools and mediating tools (Ihde, 1990, 2009a; Tripathi, 2011, 2016c).

My approach in developing a "philosophy of technological culture" is inspired by Ihde's postphenomenology, which not only allows for a new answer to the classical questions in philosophy of technology, but also for a new elaboration of the perspective behind these questions. A new interpretation of phenomenology offers possibilities *par excellence* for formulating a philosophy of technology from the perspective of things (Verbeek, 2005a). Postphenomenology aims to revitalize the phenomenological tradition in a way that overcomes the problems of classical phenomenology. As Verbeek (2005a) explains the fact that classical phenomenology failed to take the locality and context dependence of human knowledge into account is understandable when the context in which it developed is taken into account (pp. 106–108). Phenomenology presented itself as a philosophical method that sought to describe "reality itself," since it opposed itself to the absolutization of the positivistic view of the world arising from modern natural science, which claims to describe reality as it actually is. But the way in which phenomenology proceeded to develop its alternative to science, did not in fact result in a competing way of describing reality, but rather an analysis of the relations between humans and reality, explains Verbeek (2007b).

PHILOSOPHY OF TECHNOLOGY

The first explicit title which employs the term "philosophy of technology" comes from the 1877 book published by Ernst Kapp (1808–1896), *Grundlinien einer Philosophie der Technik*. Kapp was one of the century "left Hegelians," as was his slightly younger peer, Karl Marx (1818–1883), who stood at the origin of what was to become philosophy of technology as a thematic, sub-disciplinary field within philosophy (Ihde, 2004a).

Ihde (2004a) maintains that both Kapp and Marx were "materialist" Hegelians and thus technologies become much more thematic in both their developments. Kapp took "colonization" as a major metaphor and argued that there is both an "internal" and an "external" colonization which characterized the human development of technologies (p. 93).

Based on Kapp's theory of human extension, Ihde (2004a, p. 94) argues, technologies were analogized as extensions and magnifications of human organic processes and projected into an external environment. While the analogies to muscle power for various tools clearly extend specific human powers, Kapp also analogized technologies into vaster and more complex systems—for example, communications systems are analogized upon the nervous system, etc. In some sense, then, Ihde claims, technologies are material extensions of human embodiment (Ihde, 2004a).

But, why did humans make an axe? Because their hands were too weak to chop the wood. Why did people come up with the idea for a spear? Because their arms were too short and their legs were too slow to catch a running animal. Why did people invent lenses? That was because their eyes were not capable of seeing very small things, or things that were very far away. Likewise, all technical artifacts such as laptops and pens can be explained to be extensions of human body. Kapp's theory of extension of human body seems quite plausible. However, as technologies get more complex and multifaceted, it is more difficult to see in what sense they are extensions of our human bodies. Instruments tell us about the inadequacies of the human body (Tripathi, 2015, 2016a; Verbeek, 2015).

As technologies get more complex, it is more difficult to see in what sense they are extensions of our human functions. Let us take the Internet as an example. In a way, this can be seen as an extension of our human voice, because it replaces us telling the information to each other. But the system is very complex and contains several elements that are not directly extensions of the human voice and its effects are much more than just extending the amount of shared information over larger distances (can technologies here be considered as culturally multistable? Or have we multistable hermeneutic relations with the technologies?). So Ernst Kapp's analysis is too simplistic to serve as an adequate description of what the Internet is.

POSTPHENOMENOLOGY, PRAXIS, AND THE EMPIRICAL TURN

Historically, the term "postphenomenology" is introduced by Ihde (2009a) to signify a revised but thoroughly phenomenological approach to technologies and material culture; it is phenomenology applied to the study of concrete human meaning-making practices, particularly to technologies. "Classical" phenomenology—first with Husserl, including most post-Husserlians, dealt with *intentionality* (human meaning-making) but was little interested in the *technological tools* of meaning-making practices. Postphenomenology focuses on how human-technological devices affect intentionality through meaning-making practices. But rather than looking broadly at technology in general—as did 20th-century thinkers, including Heidegger—instead, postphenomenology scrutinizes particular technologies. Once philosophy of technology reached its late 20th-century state, it had become obvious that *praxis* oriented philosophies such as phenomenology or pragmatism were better suited than logic- or theory-centered analytical approaches to study the cultural and socio-historical effects of technological transformation (Ihde, 2009a).

In other words, postphenomenological research is influenced by the late 20th-century emergence of the broad set of new technoscience studies: anthropological, sociological, and feminist accounts, including actor-network theories, are especially relevant. Coming out of this set of related social science and humanities perspectives, a new gestalt concerning scientific culture has emerged. In this context, science is now depicted as a cultural field that establishes its interpretative norms by means of elaborate practices of technological hermeneutics. Here, too, praxis analysis—particularly with respect to unpacking the complexities of laboratory instrumentation—dominates (Rosenberger & Verbeek, 2015).

Postphenomenology incorporates strands of pragmatism with its distinctive experience-praxis orientation on one side, and similarly incorporates the "empirical turn" of the science or technoscience studies disciplines, on the other. While the development of postphenomenology (Ihde, 2009a) has largely taken place within the context of contemporary philosophy of technology, it has also drawn upon pragmatism on the one side and upon recent technoscience studies (Rosenberger & Verbeek, 2015).

Tripathi (2017) elaborates on the idea of postphenomenology. Postphenomenology does not abolish the distinction between humans and nonhumans, but shows their fundamental connectedness and interrelatedness (p. 141). "Artifacts mediate ways of existence (subjectivities) and experienced realities (objectivities) not because people told them to do so, but because of the relation between humans and the world that comes about through them," explains Verbeek (2005b, p. 140). Mark Coeckelbergh elaborates:

> Contemporary philosophy of technology emphasizes this "use" aspect of technologies: what matters for understanding and evaluating technologies is not only the material artifact and its immediate and intended function, but also its unintended effects as it is and becomes part of a social and natural environment. Use varies according to who uses it, but also more generally it varies according to the social context. (Coeckelbergh, 2015)

For postphenomenology, Ihde claims that with the replacement of the "subject" by embodiment, one changes the body/mind problem in early modern philosophy into a body/body problem. Merleau-Ponty, in his works, drew his distinction between the "objectively" constituted body, the mechanical and third-person constituted body of the Cartesian sciences and the *corps vecu* or lived body as experiencing body. This is the body-in-action, outside itself already in a world. Living, my body is simultaneously and yet experientially being both inside and outside (Ihde, 2003b, 2003c). "Postphenomenology is the practical study of the relations between humans and technologies from which human subjectivities emerge, as well as meaningful worlds," explain Robert Rosenberger and Peter-Paul Verbeek (2015, p. 12). Postphenomenology is an investigation of the relationships between global culture and technology (Rosenberger, 2017; Tripathi, 2015). Cathrine Hasse extends this argument, "Postphenomenology preserves the very structure of science and technology as multistable whereas claims for objective hermeneutics depend on the physical situatedness of the body" (2008, p. 57).

Postphenomenology takes an empirical focus upon human-technology relations. But it does so with meticulous scrutiny of particular concrete forms of technologies, rather than technology in general as was the case earlier in the 20th century (Tripathi, 2016a, p. 238). By exploring postphenomenology, Ihde (2009a) addresses the cultural role of technologies in relations to perception, multiculturalism, and technoscience, and gives special consideration to the impact of image technologies, such as television and cinema, upon the contemporary world. However, "postphenomenological mediation should also consider the issues of 'cultural variability.' Postphenomenologists must address the meaning of 'socio-cultural activity.' In addition, postphenomenologists should mull over whether embodiment itself has cultural & historical variations" (Tripathi, 2017, p. 140).

Ihde (2012b) rejects the entire Cartesian tradition, and instead opts for situated knowledge, but with an interrelational ontology (p. 370). As can be seen, in each set of human-technology relations, the model is that of an interrelational ontology (Ihde, 2009a). Ihde claims:

> This style of ontology carries with it a number of implications, including ones which imply that there is a co-constitution of, for example, humans and their technologies. Technologies transform our experience of the world and our perceptions and interpretations of our world, and we, in turn, become transformed in this process. Transformations are non-neutral. And it is here that histories and any empirical turn become effectively ontologically important. This, in turn, returns us to the pragmatist insight that histories are also important in any philosophical analysis as such. (2009a, p. 44)

By elaborating Ihde's stand on postphenomenology, Tripathi (2016a) tells us about "the importance of phenomenology in postmodern era, which diverges from classical phenomenology, such as in its focus on technological mediation, its reliance on 'case studies' more familiar to the field of science studies and its kinship with many of the ontological commitments of American pragmatism" (p. 238).

Postphenomenology, in a complementary role with other science studies disciplines, remains within the trajectory of those theories which reject early modern epistemology and metaphysics, including rejection of the *subject-object* distinction, and holds, instead, to an inter-relational, co-constitutive ontology (Ihde, 2012b, p. 369). The ontology of postphenomenology is interrelational (Ihde, 2012b). Further Ihde (2012b) shows that postphenomenology attempts to move farther from early modern epistemology by, as it were, substituting *embodiment* for *subjectivity*. What postphenomenology in relation to technoscience focuses upon is the way the human in science *praxis* embodies instruments (p. 370).

However on the other hand, Zwier, Blok & Lemmens (2016) provides a phenomenological analysis of *postphenomenological philosophy of technology*. While acknowledging that the results of its analyses are to be recognized as original, insightful, and valuable, we will argue that in its execution of the empirical turn, postphenomenology forfeits a phenomenological dimension of questioning. Furthermore, by contrasting the postphenomenological method with Heidegger's understanding of phenomenology, as developed in his early Freiburg lectures and in *Being and Time*, it is demonstrated how the postphenomenological method must be understood as mediation theory, which adheres to what Heidegger calls the theoretical attitude. "This leaves undiscussed how mediation theory about ontic beings (i.e., technologies) involves a specific ontological mode of relating to beings, whereas consideration of this mode is precisely the concern of phenomenology" (Zwier, Blok, & Lemmens, 2016, p. 313). Further, Zwier, Blok & Lemmens explain:

> Postphenomenology unmistakably belongs to the philosophies of technology that embrace what Hans Achterhuis (2001) has dubbed the empirical turn. This implies a critical distance towards accounts in which technology appears as a singular overarching process and instead investigates technological dynamics on a micro-scale. Postphenomenology aims to empirically analyze how particular technologies as the things themselves mediate the relation between humans and their world. This has given rise to numerous analyses and detailed descriptions of how human existence is deeply and polymorphously interwoven with artifacts. (2016, p. 314)

VARIATION AND MULTISTABILITY

Postphenomenology can be characterized as a form of analysis that retains core tactical methods and emphases that were first elaborated by Husserl and

Merleau-Ponty. For example, variational analysis, particularly taken in a radically concrete or empirical form, remains central to the descriptive practice of postphenomenology. However, the result of such analyses has also led to a transformation of the earlier notions of "essences"—which, in their Husserlian guise, retained vestiges of ideality and reductionism and which proved incapable of accounting for multistable phenomena. Ihde retains the notion of "intentionality" from classical phenomenology, as he takes it to be much more like the Deweyan concept of a contextual, interrelational process akin to an ecological organism-environment notion, or, alternatively, a relativistic situation in physics whereby the relativity between positions must be taken into account (see *Technology and the Lifeworld*; Ihde, 1990).

Variational analyses provide the methodological style of this approach. With technologies, there are multiple ways in which any single technology may be related to users and multiple ways in which each technology is culturally embedded. Variations must also be considered with respect to the complex dimensions which are included in all such phenomena. Variational analysis—more precisely, the study of group-theoretic invariants—provides a rigorous method not found in early pragmatism. Thus postphenomenology can be seen as an adaptation to the late 20th- to early 21st-century philosophic needs and issues, particularly in the context of technoscience and material culture (Ihde, 2009a). Ihde (2016) places the phenomenological practice of using variations as the primary "tool" for analysis and claims "such a variational practice works well to establish both the richness of variety provided in lifeworld experience, and to locate whatever structural features may be found" (p. 111).

The very structure of technologies is multistable, with respect to uses and to cultural embeddedness in the activities producing artifacts. When learned experiences differ, so do material hermeneutics. In these processes boundaries expand and close as imaging equipment translates invisible things into visible. All knowledge possible for us implies our own situated embodiment of knowledge. The postphenomenological approach explores how bodies, including their limits and contingencies, are reflexively implied in this process (Hasse, 2008, pp. 58–59).

Ihde (2003a) uses the term "multistability" to refer to perceptual variations which exceed the usually noted bivariational ambiguities, such as Wittgensteinian "duck-rabbits," "face-vase" alternations and the like. The diverse technologies of communication and information have stimulated much speculation concerning "virtuality," the spatiality of "cyberspace," and the role of phenomena such as "avatars." By drawing from a phenomenology of multistability and some earlier examples in technological history, Ihde (2003a) showed how the mulstability structures impact upon some of the present fantasies surrounding "virtuality" and "cyberspace." Internet, video games, and other popular technologies affect the transformations of experience through the mediation of the technologies.

Ihde introduced the idea of multistability into the philosophy of technology in technology and the lifeworld. The idea is that, like a Nekker cube illusion, which can be interpretatively seen with either the top right or bottom left corner as the frontmost surface, the path that a technology will take, both in its development and in the way that it suggests interpretive possibilities for economic and political application, is neither wholly flexible nor wholly determined (Ihde, 2012a). However, Andrew Feenberg stresses the need to grasp the multistability of possible trajectories for technology as the prerequisite for democratic steering of technological development. It is only when one has grasped both the way in which a technology does indeed commit humanity to a given future and also the way in which that future is open to a limited number of possible manifestations that one can truly grapple with technologically significant decisions in a democratic manner (Thompson, 2008). Robert Rosenberger (2015) illustrates:

> Technologies such as pacemakers and neuro-stimulators have the potential to disrupt the typical postphenomenological conception of human-technology relations because this perspective so far has focused mainly on how users engage with devices through bodily interaction and perception . . . . Multistability refers to the capacity of any technology to support a variety of uses and meanings. This idea is often wielded by postphenomenologists against totalizing accounts of technology that fail to recognize the variability and context-dependency of human-technology relations . . . . In all of these cases, the technology plays a "mediating" role in human experience, coming between the user and the world, and transforming them both in the process elaborates. (pp. 130–133)

Further Rosenberger illustrates that, a central feature of the postphenomenological framework is the notion of "multistability" (2015, p. 137). The notion of multistability is first developed in Ihde's early work on human perceptual experience; there, Ihde uses Husserlian variational theory to analyze the experience of a series of visual illusions, that is, drawings that can be perceived in more than one way (1986, p. 138). When technology is introduced, both the human experiencer and the thing experienced are transformed. "At this interrelational level, once technologically mediated, technologies may be more than just another object in the world of which the human experiencer is conscious—a point with which Husserl perhaps would have agreed," argues Michael Kelly (2015, p. 508).

The development of postphenomenology has largely taken place within the context of contemporary philosophy of technology, it has also drawn upon pragmatism on the one side and upon recent technoscience studies on the other (Ihde, 2009a). A multistable technology is a structure that follows different stable trajectories that lead to variations in the artifact as it is embedded in what is termed "life worlds" in postphenomenology "collective activities" in cultural-historical theory. As Cathrine Hasse explains:

> The "same" technology takes quite different shapes in different contexts' (Ihde, 2007, p. 13). . . . This multistability becomes apparent when artefacts move between cultural spaces or when human imagination finds new uses for already established meanings of artefacts. This is apparent in, for example, Robert Rosenberger's (2012) study of how an urban subculture finds new uses for handrails by sliding on them. Even though structures limit the sets of variations, "all technologies display ambiguous, multistable possibilities" (Ihde, 2002, p. 106). As noted by Robert Rosenberger (2011, 2012), the post-phenomenological concept of multistability has two interrelated meanings: technologies are stabilised and also have multiple existences when they cross time, space and bodily positions. (Hasse, 2013, p. 88)

MACRO-PERCEPTION AND CULTURE

Technologies are culturally multistable and perception is a hermeneutical act that can be combined to form a hypothesis of multistability of human perception. My notion of multistability of human perception is based on the fact that I can interpret the perception which is a hermeneutical act, in different ways, to cope with realities in a culture. There must be more than one way of interpreting this perception, and the same can be said of the multistable hermeneutics relation with technologies. There must be different ways to interpret the technologies where technologies are culturally embedded. My notion of multistable hermeneutics relations with technologies is that we are open to interpret the technologies in different ways.

In fact, it is important to explore multistability of human perception in philosophy of technology and multistable hermeneutic relations with technologies. A multistable hermeneutic relation to technology would be one in which a user can maintain multiple stable perceptual relationships with a "readable" device. It is also to understand the irony of cultural variants of technologies (Rosenberger, 2012, 2013a).

Rosenberger (2013b) brilliantly describes Ihde's (1999) multistability approach to the character of technology to enhance the potential to mediate experience in multiple ways, its potential to possess multiple meanings for different people *in different cultures* (italics are mine).

Ihde (2004b) argues, perceptions are bodily activities, not the actions of some homunculus inside a *camera obscura* box looking at mental images which represent something "out there." In phenomenology, variations are the means by which *possibility structures* are discovered. "Phenomena may undergo variations, reversibilities and multiple possibilities" (Ihde, 2004b).

In fact, to understand the relationship between technology and culture, it is important to explore the human-technology relations. However, the multistability of cultural relations to the world implies not only that artifacts can have different meanings in different cultural contexts, but also that the same goals can be technologically realized in different ways. Ihde's favorite example, as summarized by Verbeek, "is: the difference between western navigational techniques and the traditional navigational techniques of the South Sea islanders. Western navigation is strongly instrumentally mediated and mathematical in nature—whereas South Sea islanders navigate by carefully observing stationary clouds, birds, and wave patterns" (Verbeek, 2001, p. 135). Different cultural contexts, means different "ways of seeing" can lead to the development of difficult technologies. Interwovenness makes technology multistable. Is the cultural relation to technologies multistable, or do technologies have a culture-changing power? Perceptions are mediated by technologies.

MICRO-PERCEPTION AND EMBODIMENT

Likewise, the primacy of bodily action, including embodied perception, and other forms of praxis, remains central to postphenomenological inquiry. Indeed, for the postphenomenologist, embodied considerations can often account for the variations and multistabilities of gender and cultural difference, including a cultural hermeneutics.

Embodiment, *being a body,* is also a constant within postphenomenology. But since bodies are actively perceptual and culturally-historically constituted, postphenomenology must take account of the variations and possibilities of diverse embodiments. Thus, issues of different cultures, gender, politics, and ethics are included in postphenomenological analyses; this is illustrated by Rosenberger & Verbeek (2015), who claim that postphenomenology has specialized in how forms of knowing and perceiving the world are mediated through technologies and how technologies are changing human perceptions of the world.

Verbeek (2005a) extends the work of Ihde to present an empirically rich and nuanced image of how material artifacts and technological devices shape our existence and experiences. A new interpretation of phenomenology offers possibilities *par excellence* for formulating a philosophy of technology from the perspective of things (Verbeek, 2005a). The technologies of our everyday world have one thing in common: we use them almost exclusively via the interaction with interfaces, or better referred to as "human-machine interfaces." Technologies, with their interfaces, are so deeply intergrated into our fabric of everyday life, our "coping in the world," that human beings hardly or never draw their attention to them. Human beings use the technologies in an unconscious way.

RELATIONAL ONTOLOGY

In fact, postphenomenology continues the phenomenological tradition of a "world" (inter-relationistic) ontology of objects related to one another and culturally to human subjects. In the case of technologies, for example, humans "invent" technologies; while reciprocally, technologies also "re-invent" humans. Co-constitution is recognized in a relational ontology. But, such relational ontologies are not unique to phenomenology—they are part of the family of pragmatic (e.g., organism/environment) and actor-network (e.g., humans and their nonhuman "props") ontologies as well (Ihde, 2009a).

As can also be seen, in each set of human-technology relations, the model is that of an interrelational ontology. This style of ontology carries with it a number of implications, including ones which imply that there is a co-constitution of, for example, humans and their technologies. Technologies transform our experience of the world and our perceptions and interpretations of our world, and we, in turn, become transformed in this process (Ihde, 2009a).

The central idea in the interpretation of phenomenology developed by Verbeek (2005a), demonstrates *the mutual constitution of subject and object*, or of humans and their world. Humans and their world are always interrelated. People cannot but be directed at the world around them: they are always experiencing it and it is the only place where they can realize their existence. Conversely, the world can only be what it is when humans deal with it and interpret it. In their interrelation, both the subjectivity of humans and the objectivity of their world take shape. What people are and what their world is codetermined by the relations and interactions they have with each other (Verbeek, 2003, 2005a). Technologies help to shape relations between humans and world; Ihde says "technologies can create a "technologically mediated" intentionality" a relationship between humans and world (Verbeek, 2005a, p. 116).

Rosenberger (2017) suggests that the postphenomenological conception of "multistability" (i.e., the understanding that a technology is always open to multiple uses and meanings) is especially important for the task. Rosenberger (2017) further maintains that postphenomenology and variational cross-examination can contribute helpful insights to our understanding of the social, political, and experiential situation of technology. Rosenberger is suggesting that insights from two contemporary perspectives, postphenomenology and actor-network theory are useful for drawing out the experiential, social, and political dynamics of everyday things. Rosenberger reviews and resituates several key lines a method for using them together for the evaluation of technology. As a guiding example, Rosenberger explores a paradigmatic everyday device: fire hydrants. Despite their everyday character, hydrants fulfill multiple social roles, some of them loaded with difficult and important political implications (Rosenberger, 2017).

MEDIATION

Verbeek (2005a, 2008) has built a vocabulary for understanding this mediating role of technologies. A hermeneutic and existential phenomenological direction approaches the human-world relationship from a different side. Existential phenomenology starts from "the human side." Its central question is how people realize their existence and are present in their world. Hermeneutic phenomenology starts at "the side of world," and directs itself at the ways reality can be interpreted and be present for people. Jaspers represents the existential direction in classical phenomenology, Heidegger the hermeneutic. In the postphenomenological perspective on technology, the main question is what role technological artifacts play in the interrelation between humans and their world, and in the constitution of subjectivity and objectivity that comes about in this process (Verbeek, 2005a).

Technologies mediate between people and reality and experience. Verbeek (2008) argues, "Technologies are not neutral instruments or intermediaries, but active mediators that help shape the relation between people and reality. This mediation has two directions: one pragmatic, concerning action, and the other hermeneutic, concerning interpretation" (p. 94). This phenomenon of technological mediation has two dimensions, each of them pertaining to one aspect of the relations between humans and reality (Verbeek, 2006). First, technologies help to shape how reality and experience can be present for people, by mediating human *perception* and *interpretation*. Second, technologies help to shape how humans are present in reality, by mediating human *action* and *practices*. The first dimension can be called *hermeneutic*, since it concerns meaning and interpretation; the second is *pragmatic*, since it concern human activities (Verbeek, 2005a, 2006). Technological mediation is precisely this capacity of technology, where technologies can mediate between humans and reality, by establishing specific relations between both. Ihde (2009b, p. 465) claims that "what makes technologies valuable for human practices are the non-neutral transformational capacities of these technologies. It is the subtle and profound transformation of experience." Technological mediation organizes human-world relations; in this, material hermeneutics is also playing an important role. The concept of technological mediation is significant in bringing forth the relations between humans and the world.

On the question of how to analyze the phenomenon of technological mediation, Verbeek radicalizes Ihde's phenomenological approach of technology and offers a valuable framework in his work. In their analyses, Ihde and Verbeek understand technological mediation as the role technology plays in the relation between human beings and their world. Verbeek writes that "Ihde discerns several relationships human beings can have with technological artifacts." I agree with Ihde and Verbeek that technologies can be "embodied" by their users, making it possible that a relationship comes about between humans and their world, and also technological artifacts are "incorporated" here, as it were: they become extensions of the human body (Verbeek, 2005a, 2007a, 2008).

Technologies transform our experience of the world and our perceptions and interpretations of our world, and we, in turn, become transformed in this process. Transformations are non-neutral. And it is here that histories and any empirical turn become effectively ontologically important. This, in turn, returns us to the pragmatist insight that histories are also important in any philosophical analysis as such (Ihde, 2009a). Verbeek (2006) addresses the ethical aspects of persuasive technologies. By integrating the concept of "persuasive technology" with the concept of "technological mediation," as developed in theoretical frameworks from the field of persuasive technology and from the philosophy of technology, it will be possible to identify three such points of application: human-technology-world (in relation to perception and action) (Verbeek, 2006).

A hermeneutic and existential phenomenological direction approaches the human-world relationship from a different side. Existential phenomenology starts from "the human side." Its central question is how people realize their existence and are present in their world. Hermeneutic phenomenology starts at "the side of world," and directs itself at the ways reality can be interpreted and be present for people. In the postphenomenological perspective on technology, the main question is what role technological artifacts play in the interrelation between humans and their world, and in the constitution of subjectivity and objectivity that comes about in this process (Verbeek, 2005a). A phenomenology of human-technology relations shows that the structural dimensions of technological mediation produce a range of possible experiences (Ihde, 1990, 2006).

MATERIAL HERMENEUTICS

Hermeneutics in the traditional sense has to do with understanding and the conditions for understanding a text or a person or a situation. In philosophical hermeneutics, the historical character of understanding is posited such that one *always already understands in a certain way*, and this shapes the questioning that one does. Understanding is dialogical, a dialogue of question and answer, and one moves toward reaching an understanding with the person or the text in a process of question and answer and eventually a fusion of horizon. Newer approaches to hermeneutics, as outlined by Ihde (1998) who claims that hermeneutics applies to the very praxis of science and technology use as well as to the constitution of scientific objects (Tripathi, 2015a; Friis, 2015).

Material hermeneutics is the art of deploying new variation to interpret and understand technologies. Traditionally, the hermeneutics is used to deal with the bible texts. But when we want to interpret and understand our technologically mediated lifeworld, then traditional hermeneutics have limitations. Ihde examines hermeneutics practice within the domains of technoscience and he calls as material hermeneutics. Ihde (1998) examines what he calls a new interpretation based on material practices relating to imaging technologies which have given rise to the visual hermeneutics in technoscience studies. Historically, Ihde has sought to explicate a hermeneutic convalescence of technology (Tripathi, 2016a, p. 236). *Material hermeneutics* retains the critical, interpretive work which all hermeneutics requires, but it is more a *perceptual* than a linguistic interpretation (Ihde, 2009a). In fact Ihde's aim in *Expanding Hermeneutics* is to show that science is a profoundly hermeneutic activity, and that hermeneutics, therefore, is not limited to the humanities (Verbeek, 2003).

By explicating Ihde's material hermeneutics, Verbeek demonstrates that "hermeneutics should not only direct itself at the *linguistic*, but also at the *perceptual* aspects of interpretation. Hermeneutics is the art of interpretations: it concerns the ways in which reality can be present for people" (2003, p. 92). Perception has a hermeneutic dimension and constitutes a relationship between humans and reality. *Expanding Hermeneutics* makes possible a "connection between phenomenologically oriented philosophy of technology and empirical science and technology studies. It opens many new and interesting lines of thinking about technology and its relation to science," as Verbeek further explains (2003, p. 96).

By analyzing how technologies help to shape people's experience, an "expanded hermeneutics" arises. The traditional focus of hermeneutics on texts is broadened to things. Not only ideas and meaning, but architectures, bridges, MRI scanners, and GPS as well play a role in human interpretations. This "material turn" in hermeneutics creates a new route for understanding the role technology plays in people's interpretive relation with the world (Verbeek, 2005a).

Material hermeneutics retains the critical, interpretive work which all hermeneutics requires, but it is more a *perceptual* than a linguistic interpretation (Ihde, 2009a). A material hermeneutics is a hermeneutics which "gives things voices where there had been silence, and brings to sight that which was invisible"; such a hermeneutics in natural science can best be illustrated by its imaging practices, as Ihde (2009a, pp. 63–80) explains. The objects of this visual hermeneutics were not texts nor linguistic phenomena, but things which came into vision through instrumental magnifications, allowing perception to go where it had not gone before. Visual hermeneutics is a perceptual hermeneutics with a perception which goes beyond texts (Ihde, 2009a; Friis, 2015). Ihde tried to outline a small glimpse of this direction in *Expanding Hermeneutics* (Ihde, 1998). Such material hermeneutics are doubly material—first, in the sense that the objects being investigated are material entities—paramecia, extra-geocentric satellites, and eventually even the chemical make-up of the stars-but also it is material in the sense that the instruments being used to "bring close" such phenomena are also material entities, technologies, by which and through which the natural sciences are embodied. Our instruments and technologies operate in hermeneutic ways (Ihde, 1998; Verbeek, 2003; Friis, 2017).

Hasse (2008) argues "materiality changes with perception. Perception changes with embodied learning of artifacts" (p. 58). Technological artifacts stand in relationships with human beings and these relationships can be explored through a method of variation in which almost any artifact may become what Ihde (2012a) has called multistable (Hasse, 2013, p. 87). Our embodied subjectivities as co-constituted in interaction with material artifacts. Human engagement with technology is always meaningful to humans, even when we do not explicitly reflect on the conditions of these engagements, Hasse explains (2013, p. 81).

The perception[1] is a hermeneutical act. Friis (2015) illustrates how hermeneutics can be applied to sciences and scientific interpretation: what role does hermeneutics play in the investigations of scientific discovery, and what does hermeneutics mean for scientists, who may find themselves in hermeneutic situations. As above, I mention, that perception is a hermeneutical act, similarly Crease (1997) informs us that in Heelan's observation also the perception is not an automatic act by an independent subject (i.e., free of cultural and determination), but is *hermeneutical* insofar as it is an interpretive act guided by human involvement with the perceived (i.e., by the world).

Technologies are products of human ingenuity and designed to give us images of a reality hidden from us, either in the form of being too small or too far away and/or being inside something else, these are the technological outputs (images) (see Friis, 2017).

DIGITAL HERMENEUTICS AS MATERIAL HERMENEUTICS

Technology is a form of culture or a particular kind of culture. Culture of human performance has become sedimented in habits and traditions. Technology is shaping the theoretical outline of our social reality. The technological form of life is part and parcel of culture, just as culture in the human sense unavoidably implies technologies. There are unfathomable effects of technology on human culture and society. As Terry Winograd and Fernando Flores explain in *Computers and Cognition: A New Foundation for Design*:

> All new technologies develop within the background of a tacit understanding of human nature and human work. The use of technology in turn leads to fundamental changes in what we do, and ultimately in turn what it is to be human. We encounter the deep questions of design when we recognize that in designing tools we are designing ways of being. (1986, p. xi)

According to Rafael Capurro (2010), the story of digital hermeneutics begins with the discussions dealing with artificial intelligence (AI) in the 1970s, and particularly with Hubert Dreyfus's book *What Computers Can't Do: A Critique of Artificial Reason*, where he pointed to the importance of the context of everyday practices in which we are embedded, before we start with any kind of knowledge objectivations and their symbolic manipulations in AI systems (Dreyfus, 1972, 1992). Hubert and Stuart Dreyfus discuss Husserl's problem of objectivizing the shared background of a belief system that should correlate the everyday context of the lifeworld and could be used to make computers intelligent (Capurro, 2008).

The role of computer in the world has evolved from specialized computing machines to information devices that pervade our daily lives. As research in artificial intelligence attempts to make computers more human, some approaches to human-computer interaction are becoming analogous to human-human interaction. By attempting to emulate human conversation, natural language technologies are poised to replace traditional graphical interfaces as a more natural means of interaction. This approach, however, overlooks the embodied nature of communication, leading to serious difficulties in usability and implementation. Computers that monitor and measure the affect of students in the classroom can give helpful feedback to teachers. Recognizing other peoples' emotions and feeling or being affected by them are two different things, however. How affective computer may induce the emotional context of a certain environment is an important problem to solve: "the emotions of the game change how a player sees the field, and those aren't things that one can get a feel from the film" (Coyne, 1995). The computer's intentional arc, with the addition of multimodal and affective computing, is still incomplete (Tripathi, 2014, p. 202).

It is difficult to imagine humans as reasoning beings free of all technological augmentation since, for me, reason proceeds by the use of tools (rules are tools). I am yet to be convinced that unmediated contact is possible between the human and the world. The computer and the user form a system. And in this system, the human gives mind to the machine. The question for our time is whether machines "give mind" to the human or whether machines that appear to "give mind" to humans are but mediating instances and instruments through which other humans mind humans. It is a question if our time is the time and place of Western-inspired ideological systems that place the human in a particular situation vis-à-vis the natural in an exploiter-exploited relation. The question of the autonomy of the artifact is familiar to text encoders and ethnobotanists. It is a moral and aesthetic question that is older than the fancy talk of technological ecologies and textual economies (Tripathi, 2014, p. 203).

Paul Edwards describes the cultural-ideological background based on the introduction of the computer technology. The cyborg figure defined not only a practical problem and psychological theory, but also a whole set of subjective positions. Cyborg-brains can be understood as the machine subjects, which could be reconstructed, produced, and organized. The whole series of perspectives, self-interpretation pattern brought out the social roles in the society. It concerns a mechanical robot of the *Star Wars*. The discourses around the cyborg functioned as a psychological and subjective response to the policy of the closed world (Edwards, 1996).

The term cyberspace describes an environment, which is caused by a computerized communications networks, interactive mass media and multimedia. Cyberspace is the connection of digital information and human perception. Cyberspace promises us bright and outstanding technological future, with uncontrolled communication and a reinvention of the liberals and democratic society (Dreyfus, 2001). We may produce a certain kind of *humans* in the cyberspace, which can be authentically justified. The materiality is equipped to exist their life in an artificially safe world of the computer and in the form of computer simulations, which want to escape the world of an electronically supported environment with other individuals. In the cyberspace, distance does not play a role. Cyberspace implies the concepts of space and places. At the same time, the cyberspace concerns the metaphors of communication. In addition, a new Phenomenology of the World of this Area and the Place is necessary to be examined (Tripathi, 2014, p. 203).

Hermeneutics is, as Capurro (1990, 2001, 2006, 2010) tries to show in his work, intimately related since the 1970s with digital technology. After having passed through critical theory (Jürgen Habermas), critical rationalism

(Karl Popper), analytic philosophy (Ludwig Wittgenstein), deconstructivism (Jacques Derrida), the phenomenology of the symbol (Paul Ricoeur), psychoanalysis (Jacques Lacan), dialectic materialism (Alain Badiou), mediology (Régis Debray), the hermeneutics of the subject (Michel Foucault), and particularly through Gianni Vattimo's "weak thought" (*pensiero debole*)—to mention just some prominent contemporary philosophic schools—hermeneutics is facing today the challenge arising from digital technology becoming what Capurro calls *digital hermeneutics* (2010). In his postdoctoral dissertation *Hermeneutics of Scientific Information,* published in 1986, Rafael Capurro explores the question of information retrieval as an interpretation process of bibliographic data stored in a computer.The question of relevance and pertinence in information retrieval that plays a key role in information science can thus be hermeneutically re-considered with regard to different horizons of expectations based on the Gadamerian concept of "fusion of horizons" (Capurro, 2008).

Tripathi (2016c) demonstrates that a newer approach of hermeneutics, digital hermeneutics, applies to the concrete praxis of technologies such as Internet technology and cyberspace. *Digital hermeneutics* as defined by Capurro (2010) is understood as the encounter between hermeneutics and digital technologies that is deeply rooted in material culture, argues Tripathi (2016c). The hermeneutics of technology is understood as a hermeneutics of practice in the understanding of technologies, which is culturally and socially embedded. This cannot be done with semantics; rather this digital hermeneutics as a material hermeneutics can be explored with human embodiment. Hermeneutics which is developed for the digital world contains multistable hermeneutic relations (Tripathi, 2016c).

The multistable of artifacts implies not only that artifacts can have different meaning in different contexts, but also that specific goals can be technologically realized in different ways by a range of artifacts (Verbeek, 2005a, p. 136). In fact, it is important to explore multistability of human perception in philosophy of technology and multistable hermeneutic relations with technologies. A multistable hermeneutic relation to technology would be one in which a user can maintain multiple stable perceptual relationships with a "readable" device. It is also to understand the irony of cultural variants of technologies (Rosenberger, 2012, 2013a).

Material hermeneutics is dealing with the art of embodied interpretation of material culture and technologies (Tripathi, 2016a, 2016b). When we want to interpret and understand our technologically mediated culture and lifeworld, then traditional hermeneutics show its limitations. In this perspective, Ihde has illustrated the need to expand the notion of hermeneutics, as shown by Ihde (1998) and (Verbeek, 2003). *Expanding Hermeneutics* emphasizes the cultural situatedness of human perception (Friis et al., 2012, p. 250). The main point of an expanded hermeneutics is that what the natural sciences teach us is that there are ways, through instruments and technologies by which things can show themselves. Hermeneutics in natural sciences can best be demonstrated by the imaging practices, called as visual hermeneutics, claims (Ihde, 1998, 2009a). Indeed, the most important value of *Expanding Hermeneutics*, however, lies in its contribution to the connection between the philosophy of science and the philosophy of technology (Verbeek, 2003).

In the contemporary philosophy of technology, it is imperative to explore the primacy of practice in hermeneutical pragmatism and at large it tells us that practice-immanent theorizing is one of the factors of hermeneutical pragmatism (Fairfield, 2000, p. 4; Tripathi, 2016c). New hermeneutics of technology, which is profoundly sedimented in materiality, gives us a new meaning of technology in their usage and at the same time new hermeneutics also illustrates the sense and non-sense. The reinterpretation of phenomenology is vital, since it creates the prospect for a new phenomenological philosophy of technology, which goes beyond the classical diagnosis of alienation (Verbeek, 2005; Tripathi, 2016c, 2017). Visual hermeneutics are from this perspective not a subjective reading of an external representation, but a lived embodied experience forming perceptions of the material worlds (Hasse, 2008, p. 57).

Bielby & Kelly (2016) inform us that Tripathi (2016c) uses Ihde's insights to

> inform his own pragmatic approach to cultural and technological hermeneutics and reinvigorates these themes with a call to philosophers and information scientists to engage with the embodied realm of technology, and information, and to delve deep within their traditions and their own practice to help reveal the "basic cultural and ethical conditions of technological and economic development." (Bielby & Kelly, 2016, p. 17)

Ihde (2009a) and Friis (2017) have shown *how instrumental changes affect the sciences using technologies devices*. For the contemporary world, that which had not been visible can now become visible; and that which was unheard can now begin to be heard. Things, too, have or may be given voices (Ihde, 2009a; Friis, 2015). In Ihde's work, we find that hermeneutics derives from our innate ability to interpret, that is to say, to act, and to reflect upon pre-consciously sensuous information, as well as this innate interpretative ability he calls "perceptual thinking" (Friis et al., 2012, p. 250). One could also say that a visual hermeneutics is a perceptual hermeneutics with a perception since it does not deal with texts (Ihde, 2009a, pp. 63–80; Friis, 2015). The instruments being used to "bring close" such phenomena are also material entities, technologies, by which and through which the natural sciences are embodied (Friis, 2017).

NOTE

1. I thank Jan Kyrre Berg Olsen Friis for the e-mail exchange on this in July 2015: Thinking is not something that happens in consciousness, it takes place before conscious awareness, and this form of thinking is very much an *attuning* to sensuously givens, thus thinking is a "seeing" of that which eventually will become a perception that has been informed by past experiences and habits.

REFERENCES

Achterhuis, H. (Ed.). (2001). *American Philosophy of Technology: The Empirical Turn*. Bloomington, IN: Indiana University Press.

Bielby, J., & Kelly, M. (2016). Introduction, information cultures in the digital age: A Festschrift in honor of Rafael Capurro. In M. Kelly & J. Bielby (Eds.), *Information Cultures in the Digital Age: A Festschrift in Honor of Rafael Capurro* (pp. 1–28). Wiesbaden: Springer.

Capurro, R. (1990). Ethik und Informatik. Die Herausforderung der Informatik fuer die praktische Philosophie. *Informatik-Spektrum 13*, 311–320.

———. (1995). *Leben im Informationszeitalter*. Berlin: Akademie Verlag.

———. (2001). Beitraege zu einer digitalen Ontologie. http://www.capurro.de/digont.htm

———. (2006). Towards an ontological foundation of information ethics. *Ethics and Information Technology, 8*(4), 157–186.

———. (2008). *Interpreting the Digital Human*. http://www.capurro.de/wisconsin.html

———. (2010). Digital hermeneutics: An outline. *AI & Society, 25*(1), 35–42. doi:10.1007/s00146-009–0255-9

Coeckelbergh, M. (2015). The art of living with ICTs: The ethics–aesthetics of vulnerability coping and its implications for understanding and evaluating ICT cultures. *Foundations of Science*, 1–10. doi:10.1007/ s10699–015-9436-9

Coyne, R. (1995). *Designing Information Technology in the Postmodern Age: From Method to Metaphor*. Cambridge, MA: MIT Press.

Crease, R. (1997). Hermeneutics and the natural sciences: Introduction. *Man and World, 30*, 259–270.

Dreyfus, H. L. (2001). *On the Internet: Thinking in Action*. London & New York, NY: Routledge.

Edwards, P. N. (1996). *The Closed World: Computers and the Politics of Discourse in Cold War America*. Cambridge, MA: MIT Press.

Fairfield, P. (2000). *Theorizing Praxis: Studies in the Hermeneutical Pragmatism*. Bern: Peter Lang Publisher.

Friis, J. K. B. O., Hickman, L. A., Rosenberger, R., Scharff, R. C., & Ihde, D. (Eds.). (2012). Book symposium on Don Ihde's "Expanding hermeneutics: Visualism in science." *Philosophy & Technology, 25*(2), 249–270. doi:10.1007/ s13347-011-0060-5

Friis, J. K. B. O. (2012). Interpreting the visual. In J. K. B. O. Friis, L. A. Hickman, R. Rosenberger, R. C. Scharff, & D. Ihde (Eds.), Book symposium on Don Ihde's "Expanding hermeneutics: Visualism in science" (pp. 249–253). *Philosophy & Technology, 25*(2), 249–270. doi:10.1007/s13347-011-0060-5

Friis, J. K. B. O. (2015). Towards a hermeneutics of unveiling. In R. Rosenberger & P.-P. Verbeek (Eds.), *Postphenomenological Investigations: Essays on Human-Technology Relations* (pp. 215–225). Lanham, MD: Lexington Books/Rowman Littlefield Press.

———. (2017). Gestalt descriptions embodiments and medical image interpretation. *AI & Society, 32*, 209–218. doi:10.1007/s00146-015-0615-6

Hasse, C. (2008). Postphenomenology: Learning cultural perception in science. *Human Studies, 31*(1), 43–61. doi:10.1007/s10746-007-9075-4

———. (2013). Artefacts that talk: Mediating technologies as multistable signs and tools. *Subjectivity, 6*, 79–100. doi:10.1057/sub.2012.29

Ihde, D. (1986). *Consequences of Phenomenology.* Albany, NY: SUNY Press.

———. (1990). *Technology and the Lifeworld: From Garden to Earth.* Bloomington, IN: Indiana University Press.

———. (1993). *Postphenomenology: Essays in the Postmodern Context.* Evanston, IL: Northwestern University Press.

———. (1998). *Expanding Hermeneutics: Visualism in Science.* Evanston, IL: Northwestern University Press.

———. (1999). Technology and prognostic predicaments. *AI & Society, 13*, 44–51.

———. (2002). *Bodies in Technology.* Minneapolis, MN & London: University of Minnesota Press.

———. (2003a). Multistablity in cyberspace. In M. Hård, A. Lösch, & D. Verdicchio (Eds.), Proceedings from *Transforming Spaces: The Topological Turn in Technology Studies* (pp. 1–9). Darmstadt. http://www.ifs.tu-darmstadt.de/fileadmin/gradkoll//Publikationen/space-folder/pdf/Ihde.pdf

———. (2003b). If phenomenology is an albatross, is post-phenomenology possible? In D. Ihde & E. Selinger (Eds.), *Chasing Technoscience: Matrix for Materiality* (pp. 131–146). Indianapolis, IN: Indiana University Press.

———. (2003c, August 7). Beyond the skin-bag: The brain extends its reach outside the body, so are we all cyborgs? Review of "Natural-born cyborgs: Minds, technologies, and the future of human intelligence." *Nature, 424.* www.nature.com/nature

———. (2004a). Philosophy of technology. In P. Kemp (Ed.), *World and worldhood / Monde et mondaneite* (pp. 91–108). Dordrecht: Springer.

———. (2004b). The ultimate phenomenological reduction. In S. Arakawa & M. Gins (Eds.), *Interfaces: Image Text Language*, no. 21/22 (pp. 59–68). Worcester, MA: College of the Holy Cross.

———. (2006). Postphenomenology and the lifeworld. In M. Geib (Ed.), *Phenomenology and Ecology* (pp. 39–51). Pittsburgh, PA: Simon Silverman Phenomenology Center.

———. (2009a). *Postphenomenology and technoscience: The Peking University lectures.* Albany, NY: SUNY Press.

———. (2009b). From da Vinci to CAD and beyond. *Synthese, 168,* 453–467. doi:10.1007/s11229–008-9445-0

———. (2012a). *Experimental Phenomenology (2nd edition): Multistabilities.* Albany, NY: SUNY Press.

———. (2012b). 'Cartesianism' redux or situated knowledges. *Foundations of Science, 17,* 369–372. doi: 10.1007/s10699–011-9243-x

———. (2016). *Husserl's Missing Technologies.* New York, NY: Fordham University Press.

Kelly, M. R. (2015). Edmund Husserl. In J. B. Holbrook & C. Mitcham (Eds.), *Ethics, Science, Technology, and Engineering: A Global Resource* (pp. 507–509). Farmington Hills, MI: Macmillan Reference USA.

Peters, M. A. (2006). Towards a philosophy of technology in education: Mapping the field. In J. Weiss, J. Nolan, J. Hunsinger, & P. Trifonas (Eds.), *The International Handbook of Virtual Learning Environments* (pp. 95–116). Springer.

Rosenberger, R. (2008). Perceiving other planets: Bodily experience, interpretation, and the Mars Orbiter camera. *Human Studies, 31*(1), 63–75. doi:10.1007/s10746-007-9078-1

———. (2011). A case study in the applied philosophy of imaging: The synaptic vesicle debate. *Science, Technology, & Human Values*, *36*(6), 6–32.

———. (2012). The body as image interpreter. In J. K. B. O. Friis, L. A. Hickman, R. Rosenberger, R. C. Scharff, & D. Ihde (Eds.), Book symposium on Don Ihde's "Expanding hermeneutics: Visualism in science" (pp. 257–261). *Philosophy & Technology, 25*(2), 249–270. doi:10.1007/s13347-011-0060-5

———. (2013a). Mediating Mars: Perceptual experience and scientific imaging technologies. *Foundations of Science*, *18,* 75–91.

———. (2013b). The importance of generalized bodily habits for a future world of ubiquitous computing. *AI & Society, 28,* 289–296.

———. (2015). Postphenomenology: What's new? What's next*?* In J. K. B. O. Friis & R. E. Crease (Eds.), *Technoscience and Postphenomenology: The Manhattan Papers* (pp. 129–148). Lanham, MD: Lexington Books.

———. (2017). On the hermeneutics of everyday things: Or, the philosophy of fire hydrants. *AI & Society, 32,* 233–241. doi:10.1007/s00146-016-0674-3

Rosenberger, R., & Verbeek, P.-P. (2015). A field guide to postphenomenology. In R. Rosenberger & P.-P. Verbeek (Eds.), *Postphenomenological Investigations: Essays on Human-Technology Relations* (pp. 9–40). Lanham, MD: Lexington Books/Rowman Littlefield Press.

Thompson, P. B. (2008). The agricultural ethics of biofuels: A first look. *Journal of Agricultural and Environmental Ethics*, *21*, 183–198. doi:10.1007/s10806-007-9073-6

Tripathi, A. K. (2011). Paradigmatic shifts in the contemporary philosophy of technologies: Culture of technological reflections. In A. K. Tripathi (Ed.), *Bernhard Irrgang: Critics of Technological Lifeworld. Collection of Philosophical Essays* (pp. 13–45). Frankfurt a.o: Peter Lang.

———. (2014). Philosophy of technology reflection. *International Journal of Applied Research on Information Technology and Computing*, *5*(3), 201–213. doi : 10.5958/0975-8089.2014.00012.8
———. (2015). Postphenomenological investigations of technological experience. *AI & Society*, *30*(2), 199–205. doi: 10.1007/s00146-014-0575-2
———. (2016a). Culture of sedimentation in the human–technology interaction. *AI & Society*, *31*(2), 233–242, doi: 10.1007/s00146-015-0581-z
———. (2016b). Erratum to: Culture of sedimentation in the human–technology interaction. *AI & Society*, *31*, 243–244. doi:10.1007/s00146-015-0584-9
———. (2016c). The significance of digital hermeneutics for the philosophy of technology. In M. Kelly & J. Bielby (Eds.), *Information Cultures in the Digital Age: A Festschrift in Honor of Rafael Capurro* (pp. 143–157). Wiesbaden: Springer.
———. (2017). Hermeneutics of technological culture. *AI & Society*, *32*, 137–148. doi:10.1007/s00146-017-0717-4
Verbeek, P.-P. (2001). Don Ihde: The technological lifeworld. In H. Achterhuis (Ed.), *American Philosophy of Technology: The Empirical Turn* (pp. 119–146). Bloomington, IN: Indiana University Press.
———. (2003). Material hermeneutics. *Techne: Research in Philosophy and Technology*, *6*(3), 91–96.
———. (2005a). *What Things Do: Philosophical Reflections on Technology, Agency, and Design*. University Park, PA: Pennsylvania State University Press.
———. (2005b). Artifacts and attachment: A post-script philosophy of mediation. In H. Harbers (Ed.), *Inside the Politics of Technology: Agency and Normativity in the Co-production of Technology and Society* (pp. 125–146). Amstterdam: Amsterdam University Press.
———. (2006). Persuasive technology and moral responsibility: Toward an ethical framework for persuasive technologies. *Persuasive*, *6*, 1–15.
———. (2007a). Beyond the human eye: Technological mediation and posthuman visions. In P. Kockelkoren (Ed.), *Mediated Vision* (pp. 43–53). Rotterdam: Veenman Publishers and ArtEZ Press.
———. (2007b). *The Technological Mediation of Morality: A Post-Phenomenological Approach to Moral Subjectivity and Moral Objectivity.* In proceddings from *Workshop Moral Agency and Technical Artifacts*, NIAS, 10–12 May 2007, The Hague.
———. (2008). Morality in design: Design ethics and the morality of technological artifacts. In P. E. Vermaas, P. Kroes, A. Light, & S. A. Moore (Eds.), *Philosophy and Design: From Engineering to Architecture* (pp. 91–103). Dordrecht: Springer.
———. (2015). Beyond interaction: A short introduction to mediation theory. *Interactions*, *22*(3), 26–31.
Zwier, J., Blok, V., & Lemmens, P. (2016). Phenomenology and the empirical turn: A phenomenological analysis of postphenomenology. *Philosophy & Technology*, *29*, 313–333. doi:10.1007/s13347-016-0221-7

Chapter 12

Studying the Telescopes of Others

Toward a Postphenomenological Methodology of Participant Observation

Cathrine Hasse

INTRODUCTION

In his (2016) book on Husserl, Don Ihde firmly places postphenomenology as the successor to classical phenomenology by probing into the very foundation of the philosophy of phenomenology and shaking it up. When the "father" of phenomenological methodology, Edmund Husserl, is exposed through Ihde's postphenomenological work, Husserl's missing technologies (Ihde, 2016) emerge. From the glasses on Husserl's own nose and his writing tools to his lack of attention to Galilei as a lens grinder and user of telescopes, Ihde shows that Husserl got it wrong because the technologies used in science, including Husserl's own, remained invisible to him (Ihde, 2016, p. 46). In this chapter, I will expand this argument from an anthropological and methodological point of view. The technologies we use in research are indeed often invisible to us, just as our own positioned bodies are (Leder, 1990). Doing postphenomenology involves awareness of the glasses on our own noses as well as how we ourselves engage with the technologies we study. This awareness firmly places postphenomenology within the range of the new materialist movement with a focus on nonhumans (Bruno Latour and Donna Haraway have different but compatible things to say spanning road bumps and dogs), the material turn (Karen Barad and Tim Ingold differ in their analysis; yet, both explore how the universe meets us half way and Jane Bennett adds vibrant materials) and object-oriented ontology (with proponents like Graham Harman and Timothy Morton).

In most of these studies, as well as in classical phenomenology and parts of postphenomenology, the philosophers take their own positioned bodily technology-mediated engagement with the world as an invisible point of departure. This is sometimes done with keen awareness of how this body is placed in time and space (in Ihde's own work his positioned body is always visible in the text, just as Tim Ingold also gives salience to his own bodily position in his texts). From a methodological point of view we can, however, ask how these bodies and their use of technologies produce the appearance of the phenomena they write about. Or, as Ihde (2016) puts it: Where *was* Husserl?

Today, the philosophers' bodies are often surrounded by computers, tablets, news media—as compared to the quill pens and typewriters that were formerly used (Ihde, 2016, p. 65). To make their arguments, philosophers, as well as STS scholars, draw on what they read, hear, and learn from public media (about climate changes or new technologies, for example), but they also often draw on their own experiences in their everyday lives. Donna Haraway writes about her dogs Cayenne and Roland (2003), Ihde about his digital hearings aids (e.g., Ihde, 2015), Peter-Paul Verbeek about seeing a child in an ultra sound scanner (2011). Doing postphenomenological work can involve using your own experiences as an illustration of a point you have already decided to make, which is then filtered through an analysis connecting personal experiences with the analytical discussion points found in the texts of other scholars.

One small exercise in postphenomenology, following the footsteps of Ihde, can be to read the texts of other scholars and try to detect how technologies mediate the life experiences written about by these writers and from which bodily positions they compose their illustrations. We use empirical material as illustrations to drive home certain philosophical points about humans and their relationships with their material environments. It is not always easy to detect where the empirical examples used to illuminate an argument connect to the scholars' bodily technology-mediated position, but often the writers stay safely within the range of their own lifeworld. They use texts from journals, magazines, and historical documents, and supplement with experiences from their own well-known lifeworld. Many excellent arguments build on the writers first-person experiences with these sources. Participant observation makes it possible to include how your own first-person experiences are challenged when you engage with the mediated experiences that mediates matter that matter to other people (see for instance Aagaard, this volume). This method gives rise to a particular awareness of the workings of collective processes of learning in human-machine relations. When we learn, we gradually align our own perception of material surroundings with our co-learners. Perceptions become collectively learned (Hasse, 2015). This is the difference from taking a point of departure in our own individual experiences, and those we can convincingly argue are collectively learned through participation and observation, while engaging in the practices of other people. During this process, we not only learn to perceive dogs, ultrasound scans, images, and sounds in new ways, but we also become aware of how others may perceive dogs, ultrasound scans, images, and sounds in ways that differ from our own initial perceptions.

Postphenomenology has recently seen an interest in expanding the concept of awareness, or as it is put by Robert Rosenberger "fields of awareness" (e.g., Rosenberger, 2012). This concept may be discussed in connection with what cultural-historical psychology has named *field of attention*, tied to learning as a process of transformation. The Russian psychologist Lev Vygotsky developed the idea of learning as a cultural and collective field of attention. Most notably, he develops the idea in his discussions on how children learn to free themselves from their immediate surroundings, when they learn to think about cultural tool use instead of just grasping and letting themselves be ruled by what is before the eye (Vygotsky, 1998). It is a basic process of learning to become cultural beings. We learn to handle and think with our cultural surroundings. Thus our field of perception is enhanced from what we perceive to what we perceive culturally. This is a process where normal being-in-the-world attentionality is transformed into a cultural voluntary attention. The theory of how voluntary attention develops in humans is closely connected to discussions of how their use of technology is inevitably tied to concept formation and collective meaning-making. Because humans can use thinking tools, like concepts, they are not entirely dependent on clues from the physical environment. Their field of attention can include abstract social phenomena and past and future elements in the present, and thus pass beyond what is in sight (Hasse, 2015, p. 143).

The two kinds of fields differ in their relation to time and space. Fields of awareness differs from fields of attention in so far as the former does not focus on the process of sign mediations as part and parcel of instrument mediation—and the latter does not focus on how technologies transform our perception of space.

Combining the two includes an awareness of sign mediation tied to "technological mediation" (Verbeek, 2016). This enhances the human-technology relations to become human-human-technology relations, thus emphasizing that humans are not isolated in their uses and perceptions of technology. Technologies are learned in processes that are basically driven by frictions in cultural ecologies where meaning and materials shift in taking center stage; and participant observation is the best way to capture these processes (Hasse, 2015).

The more philosophy, anthropology, psychology, and pragmatist traditions merge in postphenomenology, the better we can begin to move into unfamiliar places and explore the complex lifeworlds of others. This entails learning how technologies matter to others. As we learn, our first-person perception aligns with the perception of others. Examples can be found in Robert Rosenberger's (2014) study of subway benches which includes interviews with homeless people, and Anette Forss's study of how pap smear readers learn to detect cancer cells (2012), or my own studies of how physicist students learn to perceive the world as physicists (Hasse, 2008). These studies mark a shift from classical phenomenology. However, some methodological issues remain unquestioned in these new engaged postphenomenological studies, so let me begin with a brief exploration of classical phenomenology and its methods.

CLASSICAL PHENOMENOLOGY

I do not intend a thorough walkthrough of phenomenological methodologies, but will point to a few points of importance when studying how technologies mediate for others. Early phenomenological methodologies were overtly formed on introspection of our own experiences. Since Franz Brentano, a transitional predecessor, phenomenology has had an on-off relationship with psychology—especially with regard to first-person perception. What appears to us, Brentano claimed, comes from ourselves (1981). He had no explicit focus on materials. Brentano aimed a rigorous philosophy building on knowledge based on direct experience. Yet, like Ihde in his discussion of Husserl, it is clear that Brentano rarely included his own use of technologies in his introspections. Furthermore, he was criticized because empirical data came through an introspection of his own first-person perspective, not through the psychological experiments from a third-person point of view that later came to define psychology. For Brentano what mattered was the precision of the descriptions following introspection. A descriptive psychology differed from a genetic psychology in so far as it did not ask about *how* phenomena appeared, but rather *what* appeared. Through this methodology, Brentano explored the difference between physical and mental phenomena, and introduced the concept of intentionality. Mental phenomena, he argued, spring from "inner consciousness" and are intentionally directed at physical objects. Mental phenomena are "only perceived in inner consciousness, while in the case of physical phenomena only external perception is possible" (Brentano, 1981, p. 91).

Brentano was a singular person scrutinizing his own perception of (basically any) objects in his own lifeworld. With a non-materialist psychological approach, the ontological problem for Brentano and his followers remained how humans experience mental phenomena that correspond to physical objects in their surroundings—some of which are graspable while others are abstract, such as an emotion, a nation-state, or a fantasy figure.

Brentano and his notion of intentionality, splitting the world into an introspecting subject and an objective physical world, inspired the phenomenological work of his student Edmund Husserl. The awareness of our own experience followed by methodological principles for investigating first-person experiences remained the pivotal point in phenomenology.

Husserl contributed with variational analysis which is still used by postphenomenologists, albeit in new ways (e.g., Don Ihde, Galit Wellner, Marie-Christine Nizzi). Through a rigorous exploration of material objects in his lifeworld (like apple trees), Husserl wanted to find the objects' essential features. In postphenomenology, the method of variational analysis is built on a pragmatic, anti-essentialist "nonfoundational and nontranscendental phenomenology which makes variational theory its most important methodological strategy" (Ihde, 1993, p. 7). The empirical material for variational analysis appears in many forms. It can be drawings like in Ihde's *Experimental Phenomenology* (1977/1986), cultural media representations like in Nizzi's study of bodies in science fiction movies (2015), or historical documents and experiences like in Wellner's study of mobile phones (2015). Postphenomenological variational analysis also includes an interest in the multistability of physical objects: how variational patterns can be found in "the idea that any technology can be taken up in different meaningful ways for different purposes in different contexts" (Rosenberger, FOS), and that humans develop different solutions to the same problems (Ihde, 2016, p. 112).

The Husserlian methodological notion of *epoché* has received less attention in postphenomenological explorations of technology. Nevertheless, both this classical approach and variational analysis are important for postphenomenological studies of other people's relations to technologies. *Epoch*, in Greek, means to "hold back," or "a cessation." In phenomenology, it means a suspension of judgment where the philosopher tries to "bracket off" all cultural, social judgment of a given object to let its pure being and essence appear (Husserl, 1931). This is—as also noted by Husserl—not an easy task. Following his former teacher Brentano, he develops the notion of intentionality tied to epoch—that is, exploring how we are directed at *intentional objects*. In epoch, these objects are explored through an act of pure consciousness suspending all cultural and social normative assumptions of why we are directed at this particular object—all in order to find its pure essence.

Maurice Merleau-Ponty (2002) grounds the immaterial body of the Husserlian subject, giving it not just a body in a lifeworld, but a world-horizon where things emerge on a background of an ambiguous and indeterminate horizon, which is constantly moving in an intersubjective space. Therefore, I see Merleau-Ponty as an intermediary figure in bringing phenomenology to postphenomenology by including the intersubjectivity of mediational processes.

Phenomenology has, since Husserl, acknowledged that our everyday experiences of the lifeworld may differ with normative and cultural perceptions. Phenomenology has, as yet, not been fully exploited methodologically. How does our engagement with technologies, even the same technologies, become so varied? What is the process? It seems, as Ihde is also hinting at in his discussion on Husserl (Ihde, 2016), that learning plays an important part.

There is more to say about phenomenological methodologies—for instance, about Heidegger's contributions—but for my discussion, here, these few examples of phenomenological methodologies will do. From Brentano and Husserl to Verbeek and Ihde, phenomenology has come a long way and there have been several important methodological developments. The discussion remains of what can be gained by studying the experiences of other people and their relations to technologies in practice, and how these postphenomenological methods of study differ from classical phenomenological work.

PARTICIPANT OBSERVATION

My own answer to this question is that through participant observation we engage in learning what matters to others. I have elsewhere argued that though participant observation is informed by theories, the relevance of the theoretical field may change when we engage as learners in the empirical field (Hasse, 2015, p. 33).

The moment we begin to study people's actual practices, our understanding of their relationship with technology changes, when we learn how practices and technologies entangle each other. Postphenomenology has opened for such empirical studies (Rosenberger & Verbeek, 2015, p. 31), making it possible to explore why participant observation is so important for a philosophy of technology.

Philosophy of technology has often focused on the very general impact of technology, which has sometimes amounted to a very technophobic analysis (see, for instance, discussions on Heidegger's dystopic technology analysis). Postphenomenology makes another argument: we should follow pragmatism and look to practices (Ihde, 2016, p. 111). I add another dimension: look at *other* people's practices. Here participant observation (possibly combined with interviews) is the most apt method.

Participant observation has many advantages and some disadvantages. It is a method often tied to anthropology, where it was developed as a reaction to the information-gathering taking place in the colonies held by Western European countries. Here, the Western European regimes needed as much information as possible about the local populations they'd imposed their regimes upon; and to that end, they made use of missionaries. These missionaries were "armchair" anthropologists. They were challenged by who was later known as the father of participant observation, the Polish-British social anthropologist Bronislaw Kaspar Malinowski. During the First World War, he was more or less kept away from his institute at London School of Economics. He spent the years from 1915 to 1918 on the Trobriand Islands, where he developed field methods based on intensive participant observation—a method which he himself described like this:

> The anthropologist must relinquish his comfortable position in the long chair on the veranda of the missionary compound, Government station, or planter's bungalow, where, armed with pencil and notebook and at times with a whisky and soda, he has been accustomed to collect statements from informants, writes down stories, and fills out sheets of paper with savage texts. He must go out into the villages, and see the natives at work in gardens, on the beach, in the jungle; he must sail with them to distant sandbanks and to foreign tribes, and observe them in fishing, trading, and ceremonial overseas expeditions. Information must come to him full-flavored from his own observations of native life, and not be squeezed out of reluctant informants as a trickle of talk. . . . Open-air anthropology, as opposed to hearsay notetaking, is hard work, but it is also great fun. (Malinowski, 1954/1926, pp. 146–147)

Anthropological methodology booms with literature about what it means to do participant observation, also in new media settings, for instance. What is important for my discussion here is that participant observation entails close attention to technologies and the consequences they have when people engage with them. Anthropologists are in a constant state of cultural epoch, in that what they took for granted is constantly challenged, and assumed normativity is decomposed. Cultural epoch does not detect "essence," but cultural variation and surprises. Even a skilled anthropologist, like Colin M.Turnbull who kept coming back to study the Mbuti Pygmies in Central Africa (1983), always learns new things. Turnbull learned by participant observation, which means to follow people and sometimes engage in doing what they do, to learn what matters. That is why a good anthropologist like Turnbull does not take anything for granted. When he walked in the woods after many years among the Mbuti he noticed leaves "lying oddly on the trail" (Turnbull, 1983, p. 119). For a first time visitor, there would be noting strange about leaves on a trail in the jungle. For Turnbull, the now familiar sight of leaves that he had learned to share with the Mbuti made him aware that something was wrong. Some leaves had a hole near the end, some were pierced horizontally and some a hole in the middle. Because Turnbull became aware, he began to notice other oddities like the interval of leaf heaps. Only then he began to ask questions. It turned out that a war was coming and the leaves were a means of telling who were friends or foes. Participant observation provides us with questions we did not know we would ask. It strips us of cultural self-evidence to provide us with new normativities of others.

From a postphenomenological point of view, participant observation opens for new questions when we study actual human-technology relationships—also in our own familiar cultures. The Danish project Technucation challenged many of our assumptions about technology use and its consequences in practice (see www.technucation.dk). By following teachers and nurses around in hospitals and schools, we gained many insights that go beyond what people say they do when they engage with technologies; partly because it is really hard to put into words how your daily engagements with technologies are constantly changing, when they turn out to have unexpected consequences (Hasse & Wallace, 2014). The material artifacts—like thermometers, black boards with chalk, or tablets—that may seem familiar to the Western participant observer in Western settings gradually gain a new meaning for the researcher when s/he follows people in their practices (Hasse, 2015).

This was also the case when, in 1996, I enrolled as a physics student to follow the students around in classes and excursions to find out how and where the transformation from "student of physics" to "a physicist" took place. There is much to be said of this strenuous journey, eventually taking me from the Niels Bohr Institute in Copenhagen to CERN in Switzerland and La Sapienza in Italy. My learning experiences among physics students began with lectures on, among others, Galilei at an old observatory in the island of Zealand. Here, we were to learn about astronomy and were introduced to the red and white Brorfelde Schmidt Telescope. I went there with a group of other students: Leonardo, Jacob Z., Viggo, Jon, Ingolf, Line, Allan, and many others from my class. In all, 23 students and 2 teachers. We are told that the Schmidt telescope is now obsolete, just like the small refraction telescopes used by Galilei in professional astronomy. Once, the Schmidt telescope was the pride of the astronomy department, but now observations can be made much better at La Silla Observatory in Chile, with their many advanced digital telescopes like the ESO New Technology Telescope (NTT). This is more than just relegating the relics to the dust heap of history. It also marked an identity shift from the hands-on work, which many physicists took pride in, to the new age of computer data. For the teachers, and some of the students, the shift from the "modern" technology of refracting telescopes to the "postmodern" imaging technologies (Ihde, 2016, p. 114) had removed a romantic aura from astronomy.

Once, it was a mark of nobility to be a skillful observer managing telescopes "the hard way," one of the teachers told us. She explained how students in her time had to sit for several nights in thermal-suits belonging to the then unheated observatory, and how they felt like the Danish astronomer

Tyco Brahe, whom they were told made his observations night after night, rubbed in seal fat. Later, a small computer room was built where we sat now. It had heating and showed pictures of the CCD photographs now part of the Schmidt telescope. Some showed the rings of Saturn, which we would later analyze through the method of spectrometry. Our teacher did not conceal his opinion that, for them, a real astronomer is one who knows how to handle a proper telescope like Galilei's, which will refract the light through a series of finely ground glass lenses. They also let us try some of these, and like Ihde's guests in Vermont, we had to learn to adjust our bodies and the technology to be able to see anything at all (Ihde, 2016, p. 30). Some students were disappointed that we could not see the fantastic colorful images of nebulas and galaxies from the pictures displayed in books. Viggo complained about having to learn about these old fashioned instruments, when we know they are of no value to astronomy any longer. Jacob, on the other hand, was keenly interested. He argued that we need to know the basics before we move to more refined instruments. He added that the colorful nebulas are second order reproductions anyway (just in line with Ihde's postphenomenological work on imaging technologies).

We all had difficulty "seeing" through the refraction telescopes at first, but we learned how to adjust the flickering lights and not to let our vision through the telescope focus on objects too close to the horizon to avoid the disturbances of the Earth's atmosphere. After a while, we had all learned Galilei's obsolete technique on how to be moon gazing. The telescopes at Brorfelde mediate more than the sight of the mountains of the Moon, however. They mediate an initiation rite into the tribe of astronomers imbued with mystique and awe of the skills of what may become our founding fathers. The message from our teachers was clear: the work of a scientist is hardship and the telescopes are a reminder of the hardship you have to endure. This is part and parcel of becoming an astronomer even if most of the work today takes place in front of a computer analyzing images.

The telescope lessons in astronomy did mediate a new world for me as well as for my fellow students. My understanding of the meaning of a "star" changed as I saw the many galaxies I had taken to be stars. In astronomy class, I not only learned about the different size classes of stars and to decode their spectrum lines, I also learned that all stars are potential solar systems surrounded by exoplanets and some of these might be just like our earth, and it would be a major new contribution to physics to find such exoplanets (this was in 1996—since then many have been found). The telescope thus enhanced my perception of the starry sky even when I was not looking into the instrument. I also learned that the meaning of "knowing the instruments by heart" was part of what was recognized as a "real physicist."

LEARNING TO ALIGN

Husserl's Galilei did not have a telescope, and that led Husserl to overlook important aspects in his critique of Galilei as a mainly mathematically oriented scientist, Ihde argues (2016). Ihde shows that the telescope of Galilei was pivotal for the development of his theories and should be part of a phenomenological analysis. Participant observation of the telescopes of others teaches us something more. Galilei and Husserl lived in their respective cultural lifeworlds (ecologies), where technologies and meaning-making entangled what was kept and what was discarded.

Because I learned with the physics students, I was not the isolated subject of Brentano, studying myself introspectively. Participant observation placed me in a constant position of epoch; not in relation to finding essence, but new possible variations from a cultural point of view. Nothing can be taken for granted, as the following example will show. When we got back from the Brorfelde excursions we had to make a spectroscopy analysis of the rings of Saturn. I came to my fieldwork with the clear understanding that physics was about being exact and precise in measurements. Much like Husserl, I believed that scientists learned a rigorous and mathematical approach to the forces of nature; whereas, in fact, this learning is embedded in a wider lifeworld.

In the process of learning, positioned as a student among students, I learned many new words belonging to a student world of slang (such as *GT*, which turned out to refer to the Danish beer Gold Tuborg). I also learned new meanings of already known words connected to the world of students. One example of this is the word "exact" which was used by the teacher and textbooks just as I expected it to be. They told us to be "exact" in our measurements. What participant observation does is to challenge and sometimes strip you of normative assumptions to let you see things as they really are when people engage with technologies and each other.

This was the case when, sometime after our excursion to Brorfelde, we were asked to analyze the rings of Saturn using Doppler spectroscopy. This is a method for detecting variation in the color of the light emanating from space objects in movement. As the rings move closer towards Earth, the spectrum of the rings are turning bluish; and as they move away, the specter turns reddish following their respective wavelengths. Using a ruler, we were to measure by millimetres the variations found in the Saturn spectrogram, an image constituted from a diameter cut through Saturn and its rings—where the rings appeared as thin bands on either side of the wider middle band of the planet itself. As I ventured into actually making exercises, I found that from this position among the students, "exact" and "measurement" carried a new meaning. If I had just looked at a movie of our measurement class exercise I would have noticed a group of students occupied with measuring, occasionally speaking a little or moving their bodies in the seats. Here I actually participated, which gave a new meaning to the spectroscopy exercise. We were sitting in the classroom side by side, each with a piece of paper measuring and calculating individually. In an individual exercise, each student was asked to measure as precisely as possible with our primitive rulers the distance between dark lines symbolizing the variations of the rings.

What I gained from this participant observation was not a spectator's knowledge, but a pragmatic knowledge fraught with understanding of how measuring matters for a physics student. We—including researchers studying other people's lifeworlds—do not learn just by activating a physiological system, but by a continuous engagement with a world where each of our actions has consequences for the next (Dewey, 1997, p. 208). Participant observation entails learning the consequences of actions in situated practices. As an active engaged participant, we learn to pay attention as we learn to think with the new meanings of material artifacts (Hasse, 2015), and the attention we obtain becomes "a mode of participation, valuable in the degree in which it is effective. It cannot be the idle view of a spectator" (Dewey, 1997, p. 210). Thus, attention is voluntary in so far it is connected to what we want to obtain or avoid. It follows that if people have problems controlling their voluntary attention when handling technologies and human-human relations (e.g., concentrating on smartphone engagement or human-human conversation [Aagaard, 2015; Aagaard, this volume]), this problem is tied to a cultural lifeworld.

What counts as effectiveness in this particular cultural setting, using ruler-technology, is not to be taken for granted either. Rulers afford exact measurement. The attention included, however, a social relation to the technology. My position as a participant observer doing what the students did, placing my body in the same position as theirs, taught me unexpected consequences of the teacher's demand for exact measurements. As an anthropologist among physics students, I was, of course, nervous that I would not do the measurements the right way. From time to time, I let my eyes wander to the guy next to me to check my measuring up against his. I tried not to be too obvious about how much I looked at his results, as I did not want to make too much of a fool of myself. Then I noticed that the student next to me was also casting glances toward the person next to *him*—and from time to time adjusted his own results according to what he saw. To my surprise, I then noticed that not only this guy, but also other students cast sidelong glances towards each other's work, and tried to adjust the measures to what they saw or heard—either by measuring again or simply by adding some numbers if they were too far ahead. "Exact measuring" in this particular setting was not an individual but a collective task. And I, by my active engagements with the task and instruments at our disposal, became part of an evolving collective consciousness of what being a physicist entails. Later interviews confirmed, together with other episodes, that many students had a fear of appearing "stupid" in the eyes of their fellow students, and thus made an effort to adjust whatever measurements they were getting to not stand out. In other words, they aligned their exactness using the ruler, with the normative culture of physics. The demand for individual exact measuring was overruled by the demand for presenting yourself as not too far from the others' results. Later participant observations among the academic staff at physics departments showed that tendencies to social alignment could also be found among the educated physicists. Their fields of attention did not only include the technologies and what they mediated. Mediation took place in a lifeworld which, as claimed by Husserl (1970), was very much a part of their scientific activity. Their attention was aimed at a future as physicists, which should not be hampered because their measurements were out of line.

STUDYING VARIATION IN MEDIATION

Why is participant observation such an important postphenomenological approach? Postphenomenological work always works from an empirical basis. After all, most postphenomenological analysis is based on analysis of very mundane technologies and investigations of how humans relate to such technologies on a more or less generalized scale. Participant observation, however, will enhance philosophical reflection by opening for new questions pertaining to human-human-technology relationships. This focus on more than *a human experience* or *this human experience* brings new dimensions to the classical notion of epoch. Through participant observation, our assumptions about the self-evident ways humans and technologies amalgamate and mediate subject-objects relations (Rosenberger & Verbeek, 2015) are challenged, and through learning, a transformed amalgamation take place. It does not bring us to the essence, but brackets off our own normativity and opens us for learning. This is not the same naïve belief that we can willfully bracket off any cultural assumptions we have, like is sometimes claimed (see Aagaard, this volume, for a discussion on van Manen). Cultural epoch does not strip us of normativity, but transforms how we are normative. Learning in other people's lifeworlds moves us from the first-person perspective of the investigator to a first-person-within-a-new-cultural-third-person perspective. Participation entails embedding and using your own bodily presence in the lifeworld of others. This investment is not without consequences. I have elsewhere written about the terrible feeling of loneliness, and the exclusion and bullying that anthropologists have to endure while they try to find their place in other people's cultural ecologies (Hasse, 2015). What we gain is an engaged learning process, where we, to some extent, come to share other people's fields of attention as well as their fields of awareness. This may include an enhanced understanding of how technologies mediate new visions—like when a telescope opens a near-far world of moon landscapes (Ihde, 2016), or our world is reduced to the small square of a mobile phone (Rosenberger, 2012). It also includes how human-human engagements with technologies entail learning new identities, passing through initiation rites, and including past and future expectations.

The important insight of participant observation for postphenomenology is that it is of equal importance *how* we see with technology as *what* we see with technology. The relation we have with telescopes, with mobile phones and tablets, the brands we prefer, reveal something about us to other people. Participant observation is a good way to obtain insights into the non-functional aspects of technologies. The way people choose and handle technologies can have a quality of what the science historian Mario Biagioi once named *sprezzatura,* after an Italian handbook in courtship. This courtly nonchalance aims at showing, through exterior signs, that you are one of the "chosen" (Biagioli, 1993, p. 51). According to Biagioli, it was this sprezzatura that earned Galilei the heart of his protectors, the Medici family. Galilei, as also noted by Ihde, never convinced the local Jesuits of the importance of his discoveries because they refused to see moon mountains or Jupiter's moons as anything other than specks in the lenses. The Medici, however, fell for his witty and eloquent presentations of the telescopic way of perception; both because it made visible what was far away in their mundane world of entertainment, and because it displayed the mountains of the Moon. As a "marvellous trick," the telescope, and with it Galilei, gained an identity as a real courtier having sprezzatura. Many new technologies today seem imbued with an ability to mediate just that kind of right identity that is going beyond functionality to tell stories about who we are, should be, and want to be. Jacob places himself as one of the future-physicists-to-be through his acknowledgment of the hardship tied to using refraction telescopes (and knowledge of imaging technologies). He possesses this quality of sprezzatura, whereas Viggo does not. Participant observation can pick up such fine-grained insights. This, again, enhances our capability to do the variational analysis that Ihde has deemed the pivotal landmark of postphenomenological analysis (Ihde, 2016, p. 111).

Placing my own body among physics students was not a static relation with technologies. As Rosenberger notes "the deeply situated character of any given postphenomenological analysis" (2014, p. 385), participant observation makes a difference in so far as it, for all its problems, opens for a situatedness of what matters in other people's lifeworlds. Telescopes never had a "thing-ness," like Ihde notes they had for Husserl (Ihde, 2016, p. 25), but were always tied to local and specific practices that transformed *what* and *how* technologies mediate. The telescopes at Brorfelde mediated not just a "near-far" world of the Moon, but in a very basic way, mediated the identity of being a scientist. These telescopes connected us with the past, and thereby the future, of physics. The ruler used in the Doppler-experiment mediated not just measurements of the rings of Saturn, but together with the social consciousness of "not being a fool," it also mediated ways of being an "exact" physicist. I learned and kept learning to become part of a forming collective consciousness of physics students. Though interviews revealed that teachers expected students to be exact, participant observation showed me that there is a social adjustment to how you relate to instruments.

STUDYING MULTISTABILITIES

Not everyone adjusted and aligned to the demands of the cultural ecology. Like Viggo, some students did not find astronomy as colorful as expected and left within the first year of study. Jacob Z. stayed on, not just because of his relation to the instruments as functional tools, but because he grasped their significance to the identity of a physicist. This process of inclusion and ejection from the physics institute was not just about telescopes. It was about diversity in how technology-human-human relations were stabilized for some students in ways that were not shared by others. Thus, the mediation of technologies was not a homogeneous collective process but revealed the complexity of the processes of stabilization.

This insight from participant observation can open for new questions into the processes behind stabilization processes, and thus expand the concept of multistability in postphenomenology. Multistability in postphenomenology covers several types of stabilizations. Rosenberger notes that multistability highlights two points:

> (1) multiple relations to a technology are always possible, and (2) this potential is at the same time limited by the technology's materiality, i.e., the particularities in its physical composition. In this terminology, an individual stable relation to a technology is known as a stability and also as a variation (these two terms are deployed interchangeably in this literature). (Rosenberger, 2014, p. 377)

Multistability is both the historical variation of the formation of technologies serving the same purpose (as when we can observe hundreds of solutions to making a "fishhook" in the Oxford-based museum Pitt Rivers). Camera obscura variations, and bow-under-tension that becomes a musical instrument, are only some of the examples taken up by postphenomenologists (Ihde, 2016, p. 129). These kind of sweeping time-space studies of a variety of stabilizations over time are not necessarily helped by participant observations. The method of participant observation is, however, excellent in exploring how human-human-technology relations vary with local practices—even when the technology could be seen as "the same." This is not just about using the telescope as a paperweight. We find variation in how human meaning-making is co-constitutive of technologies. The variational analysis of multistabilities is tied to the exploration of how the "same" artifact may be perceived differently. Again Ihde leads the way in experimental phenomenology, exploring, for example, how the duck-rabbit can be seen as a Martian (Ihde, 2016, p. 112). This is a single human exploring doodles experimentally. The work by Rosenberger on subway benches moves closer to a study of variation in human-human-technology relations. Here, we find the argument that "a technology can be used for multiple purposes through different contexts" (Rosenberger, 2014, p. 369), and that subway benches can be used by passengers for sitting, and by homeless people for sleeping. Anthropological studies have shown through participant observation, that for homeless people, finding sleeping places has salience, and that they and police officers have very different views of the surrounding artifacts—including subway benches (Spradley, 1970). Participant observation can make us aware that when we move from first-person perspective to third-person perspective, we find diversity in how humans and technologies entangle themselves in meaning-making processes in practice. This is the case in Rosenberger's study of subway benches. The passengers perceive the benches' potential for sitting, while the homeless perceive the benches' potential for sleeping. By taking the position of the homeless, visiting and learning from them, Rosenberger came to see benches with separate seats as "non-sleeping" benches, like a homeless person might do (Rosenberger, 2014).

CONCLUSION

In science and technology studies, the new materialist turn has increasingly made researchers focus on nonhumans—spanning animals, rocks, plants, and man-made materials—in a new vibrant vitalism, exploring a universe of relating bodies. Postphenomenology holds a special position on this new theoretical Parnassus in insisting that the first-person perspective of the researchers is taken into account. No matter how fruitful it may be to speculate about a world as it unfolds without any humans to perceive it, we need to take our relational processes of becoming, perceiving, and theorizing about things into account.

As noted by Verbeek: "In order to develop a full understanding of processes of mediation, we should not only study 'what *things* do' (cf. Verbeek, 2005) but also how *humans* give meaning to these mediations—both empirically and conceptually" (Verbeek, 2016, p. 190). These humans, I have argued, are first of all learners involved in a process ontology, changing their material-social world as they engage with it. The category of humans must, by necessity, include the theorizing researchers and their changing perceptions of their world. When researchers learn as they engage in the worlds of collective others (like physicists and other telescope users), the researchers' ontological world perceptions may change as well. Participant observation opens for studies of how technologies become multistable when perceived as meaningful by humans, whom they impact in ways that have different consequences for human practices and material stabilities. Postphenomenological studies "typically investigate how, in the relations that arise around a technology, a specific 'world' is constituted, as well as a specific 'subject'" (Rosenberger & Verbeek, 2015, p. 31). Physics students are formed by the technologies they learn to engage with, and they are formed by each other. They become a collective subject, and we, as participant observers, can follow them part of the way.

I am not proposing that postphenomenological philosophers should now share the benches of homeless people at night, nor that they should try to pose as "real" physicists. Yet, I do propose that studying how technologies mediate in varied ways also means that some kind of normative relation with technologies is formed; this occurring as a learning process takes place through engagement with technologies, which transforms the phenomena and aligns them with the lifeworld where they matter.

This learning process should entail more than just the human-technology relations, so thoroughly explored by Ihde and Verbeek. It should also include the human-human cultural diversity that can be further explored through participant observation.

REFERENCES

Aagaard, J. (2015). Drawn to distraction: A qualitative study of off-task use of educational technology. *Computers & Education*, *87*, 90–97.

Biagioli, M. (1993). *Galileo Courtier: The Practice of Science in the Culture of Absolutism*. Chicago, IL: Chicago University Press.

Brentano. F. (1981). *Psychology from an Empirical Standpoint, I-III* (M. Schättle and L. McAlister, Trans.). London: Routledge.

Dewey, J. (1997a). Theories of knowledge. In L. Menand (Ed.), *Pragmatism* (pp. 205–218). New York, NY: Random House.

———. (1997b). The need for a recovery of philosophy. In L. Menand (Ed.), *Pragmatism* (pp. 219–232). New York, NY: Random House.

Forss, A. (2012). Cells and the (imaginary) patient: The multistable practitioner-technology-cell interface in the cytology laboratory. *Medicine, Health Care and Philosophy, 15*(3), 295–308.

Haraway, D. J. (2003). *The Companion Species Manifesto: Dogs, People, and Significant Otherness*. Chicago, IL: University of Chicago Press.

Hasse, C. (2008). Postphenomenology—learning cultural perception in science. *Human Studies, 31*(1), 43–61.

———. (2015). *An Anthropology of Learning*. Doordrecht: Springer Verlag.

Husserl, E. (1931). *Ideas: General Introduction to Pure Phenomenology*. (W. R. Boyce Gibson, Trans.). London: George Allen & Unwin Ltd.

———. (1970). *The Crisis of European Sciences and Transcendental Phenomenology: An Introduction to Phenomenological Philosophy* (D. Carr, Trans.). Evanston, IL: Northwestern University Press. (Originally published 1936)

Ihde, D. (1986). *Experimental Phenomenology*. Albany, NY: SUNY Press. (Originally published 1977)

———. (1993). *Postphenomenology: Essays in the Postmodern Context*. Evanston, IL: Northwestern University Press.

———. (2015). Embodying hearing devices. In D. Ihde (Ed.), *Acoustic Technics* (pp. 71–78). Lanham: Lexington Books.

———. (2016). *Husserl's Missing Technologies*. New York, NY: Fordham University Press.

Leder, D. (1990). *The Absent Body*. Chicago, IL: University of Chicago Press.

Malinowski, B. (1954). Myth in primitive psychology. In B. Malinowski (Ed.), *Magic, Science and Religion and Other Essays* (pp. 93–148). Garden City, NY: Doubleday. (Originally published 1926)

Merleau-Ponty, M. (2002). *Phenomenology of Perception*. London: Routledge Classics. (Originally published 1945)

Nizzi, M. C. (2015). Movies and bodies: Variations of the embodied self in science-fiction techno-fantasies. In R. Rosenberger and P.-P. Verbeek. (Eds.), *Postphenomenological Investigations: Essays on Human-Technology Relations* (pp. 85–104). London: Lexington Books.

Rosenberger, R. (2012). Embodied technology and the problem of using the phone while driving. *Phenomenology and the Cognitive Sciences, 11*(1), 79–94.

———. (2014). Multistability and the agency of mundane artifacts: From speed bumps to subway benches. *Human Studies, 37*(3), 369–392.

———. (2016). Notes on a nonfoundational phenomenology of technology. FOS. (in print)

Rosenberger, R., & Verbeek, P.-P. (Eds.). (2015a). *Postphenomenological Investigations: Essays on Human-Technology Relations*. London: Lexington Books.

———. (2015b). A field guide to postphenomenology. In R. Rosenberger and P.-P. Verbeek. (Eds.), *Postphenomenological Investigations: Essays on Human-Technology Relations* (pp. 9–41). London: Lexington Books.

Spradley, J. P. (1970). *You Owe Yourself a Drunk: An Ethnography of Urban Nomads*. Boston, MA: Little, Brown.

Turnbull, C. M. (1983). *The Mbuti Pygmies: Change and Adaptation*. New York, NY: Holt, Rinehart, & Winston.

Verbeek, P.-P. (2011). Preface. In *Moralizing Technology: Understanding and Designing the Morality of Things* (pp. vii–ix). Chicago/London: The University of Chicago Press.

———. (2016). Toward a theory of technological mediation: A program for postphenomenological research. In J. K. Berg, O. Friis, & R. C. Crease (Eds.), *Technoscience and Postphenomenology: The Manhattan Papers* (pp. 189–204). London: Lexington Books.

Wallace, J. and Hasse C. (2014). Situating technological literacy in the workplace. In J. Dakers (Ed.), *New Frontiers in Technological Literacy* (pp. 153–164). London: Palgrave Macmillan.

Vygotsky, L. S. (1998). The crisis at age seven. In R. W. Rieber & J. Wollock (Eds.), *The Collected Works of L.S. Vygotsky*, (vol. 5). New York, NY: Plenum Press. (Originally published 1933)

Wellner, G. (2015). *A Postphenomenological Inquiry of Cell Phones: Genealogies, Meanings, and Becoming*. Lanham, MD: Lexington Books.

Index

List of Contributors

Jesper Aagaard is a research assistant and an assistant professor in the Department of Psychology and Behavioural Sciences at Aarhus University, Denmark. His research interests include philosophy of technology, media technologies, technohabitual agency, attention, and presence. His recent research includes a project on digital distraction, a qualitative analysis of empirical studies of media multitasking, drawing on postphenomenology and psychology.

Catherine Adams is an associate professor and associate chair of secondary education at University of Alberta, Canada. Drawing on links between phenomenology, pedagogy, and philosophy of technology and media scholarship, her research includes digital technology integration across in education; ethical and pedagogical issues involving digital media in schools; sociomaterial approaches to qualitative inquiry; computing science curriculum and pedagogy.

Ciano Aydin is a professor of philosophy, head of the Department of Philosophy, and vice dean of Education of the Faculty of Behavioural, Management and Social Sciences at University of Twente and professor of philosophy at Delft University of Technology in the Netherlands. He writes on the philosophy of technology with an interactionist perspective. His main research focuses on issues related to identity and (information) technology.

Lasse Blond is a PhD fellow at School of Communication and Culture at Aarhus University, Denmark, where he is affiliated with the Centre for Science and Technology Studies (STS) and the interdisciplinary research program Future Technology, Culture and Learning at the Danish School of

Education (DPU). He is currently working on his dissertation about the transfer of Korean social robots to Finland and Denmark.

Jan Kyrre Berg Friis is a course manager of theory of science at Copenhagen University. His main research areas are science and technology studies, philosophy of science, perception studies, hermeneutics, and cognition. His current research deals with the problem of perception in medicine, communication and diagnostization, and radiology and interpretation.

Michael Funk is a university assistant for philosophy of media and technology at University of Vienna, Austria. His main research areas include ethics, methodology, and epistemology of transdisciplinary research, like robotics and biology. He is currently working on his PhD thesis which addresses the methodology and epistemology of transdisciplinary research in social robotics and paleoanthropology.

Cathrine Hasse is a professor of cultural anthropology and learning at Aarhus University, Denmark. She heads the research program Future Technology, Culture and Learning at the Danish School of Education. She has written on posthuman learning, postphenomenology, technological literacy, cultural learning, roboethics, and organizational culture. Her recent research interests include educational technology, physics sciences, and robotics.

Moa Petersén is an associate professor in the Division of Digital Cultures and the Division of Art History and Visual Studies in the Department of Arts and Cultural Sciences at Lund University in Sweden. Her research interests include photography technology and human bodily self-understanding in relation to technology.

Michel Puech is an associate professor of philosophy at Paris-Sorbonne University, France. He has published numerous books and articles on the philosophy of technology, the concept of the "sustainable," and more broadly on new value systems. His current work focuses on the notion of modern wisdom, combining philosophy of technology, applied ethics, and Asian schools of thought.

Tobias Röhl is a postdoctoral researcher at the Department of Social Sciences, University of Siegen, Germany. He is currently working on a research project on breakdowns and disruptions in transport infrastructures. His research interests include the sociology of knowledge and education, sociomaterial practice, infrastructures, and science and technology studies.

Robert Rosenberger is an assistant professor in the School of Public Policy at the Georgia Institute of Technology, teaching the philosophy of technology and philosophy of science. In an ongoing series of papers, Rosenberger investigates the experience of the use of imaging technologies in scientific practice, with case studies in neurobiology and space science.

Kasper Schiølin is a senior research fellow with the Program of Science, Technology & Society (STS) at Harvard University, where he works on a project, funded by the Carlsberg Foundation, about new technologies and the accompanying socio-technical imaginaries of human perfection, at both a societal and an existential level.

Fernando Secomandi is an adjunct professor, coordinator of research, and vice-coordinator of the Post-Graduate Program in Design at ESDI-UERJ (Industrial Design College, Rio de Janeiro State University). His recent research interests include interface design, service design, and the role of the body in technological mediation.

Jessica Sorenson is a research assistant at the research program Future, Technology, Culture and Learning at Aarhus University, where she is involved in a project that involves ethnographic studies of robot development in Europe. She is currently researching ethical decision-making in a collaborative industrial robotics project.

Oliver Tafdrup is a PhD fellow at the research program Future, Technology, Culture and Learning at Aarhus University. His research interests include studies of science and technology, IT and media, philosophy of technology, educational technologies, and phenomenology. He is currently researching the use of social robots in education.

Arun Tripathi is an independent research scholar at Central University of Tibetan Studies in Sarnath, Varanasi, India. His main research areas are philosophy of technology, intercultural perception, postphenomenology of technological mediation, technology transfer, interface of human cognition and technology, and pragmatism and hermeneutics.

Joni Turville is a PhD candidate at University of Alberta and works in teacher professional development across Alberta. Her research interests are diverse, ranging from educational technology to inclusive education. Her current research examines the impact of digital technologies on pedagogical relationships and communication in K-12 schools.

www.ingramcontent.com/pod-product-compliance
Ingram Content Group UK Ltd.
Pitfield, Milton Keynes, MK11 3LW, UK
UKHW020218290926
15470UKWH00001B/263

9 781498 545235